# Petroleum Engineering

**Editor-in-Chief**

Gbenga Oluyemi, Robert Gordon University, Aberdeen, Aberdeenshire, UK

**Series Editors**

Amirmasoud Kalantari-Dahaghi, Department of Petroleum Engineering, West Virginia University, Morgantown, WV, USA

Alireza Shahkarami, Department of Engineering, Saint Francis University, Loretto, PA, USA

Martin Fernø, Department of Physics and Technology, University of Bergen, Bergen, Norway

The Springer series in Petroleum Engineering promotes and expedites the dissemination of new research results and tutorial views in the field of exploration and production. The series contains monographs, lecture notes, and edited volumes. The subject focus is on upstream petroleum engineering, and coverage extends to all theoretical and applied aspects of the field. Material on traditional drilling and more modern methods such as fracking is of interest, as are topics including but not limited to:

- Exploration
- Formation evaluation (well logging)
- Drilling
- Economics
- Reservoir simulation
- Reservoir engineering
- Well engineering
- Artificial lift systems
- Facilities engineering

Contributions to the series can be made by submitting a proposal to the responsible publisher, Anthony Doyle at anthony.doyle@springer.com or the Academic Series Editor, Dr. Gbenga Oluyemi g.f.oluyemi@rgu.ac.uk.

More information about this series at http://www.springer.com/series/15095

Martin J. Vilela · Gbenga F. Oluyemi

# Value of Information and Flexibility

## Making Decisions Under Uncertainties

Martin J. Vilela
Aura Energy Solutions
Madrid, Spain

Gbenga F. Oluyemi
School of Engineering
Robert Gordon University
Aberdeen, UK

ISSN 2366-2646                      ISSN 2366-2654  (electronic)
Petroleum Engineering
ISBN 978-3-030-86991-5              ISBN 978-3-030-86989-2  (eBook)
https://doi.org/10.1007/978-3-030-86989-2

This Springer imprint is published by the registered company Springer Nature Switzerland AG
The registered company address is: Gewerbestrasse 11, 6330 Cham, Switzerland

*To*

*My love, Aura Alvarado, for her permanent support and enthusiasm*

*To my son, Martin Antonio*

*Martin J. Vilela*

*To*

*My wife, Folasade Oluyemi, for her prayers, love, and support*

*All my teachers and others who have inspired me in learning*

*Gbenga F. Oluyemi*

# Preface

This book explores the fascinating concept of decision making in a rational and consistent manner under conditions of uncertainty to achieve what matters to the decision-makers, regardless of the choices they have.

The most important task for managers and leaders is to make decisions. In making decisions, managers need the support of their technical staff to evaluate and assess those decisions. In this sense, in one way or another, we are all involved in the decision-making process, and for this reason, it is critical to understand its foundations.

Decisions under uncertainty are part of our daily personal and professional activities. Indeed, we make decisions when selecting what to wear when leaving the house for a meeting, when or if we decide to obtain a mortgage that we need to commit to for many years or when we recommend or decide on a multi-million-dollar project that can compromise the future of our enterprise. The methodologies discussed in the book can be applied in several domains and under the circumstances outlined above.

Independently of the final outcome of a decision, the most we can do is make the decision as rationally as possible, considering all the elements related to it whilst securing consistency with our beliefs and criteria.

Even though there are several methodologies available to make decisions, we hardly use them. When they are used, we are not aware of the background underpinning them most of the time, which leads to failures due to poor compliance with the rationalisation and consistency required.

When we must decide on a project with uncertainty, we should be able to incorporate the impact of the uncertainties in its valuation, which is based on the project's outcomes that have value to the decision maker (money, happiness, health and so on). Achieving what matters to us as decision makers means we want the best possible result. To achieve this best possible result, we propose two main methodologies: the value of information and the value of flexibility.

This book covers several topics that should be part of any undergraduate or graduate programmes and provides the foundations of decision making and how decisions can be optimised.

Due to the authors' background, we mainly focus on, but are not limited to, problems related to the oil and gas industry.

The book is organised into nine chapters. Chapter 1 explores the meaning of decision making and the challenges that uncertainty brings to the decision-making process. Several standard methodologies are described, and examples are discussed.

Chapter 2 discusses the risk attitudes of several decision makers and shows how they are naturally integrated into the decision-making process. Utility functions are described, and the von Neumann and Morgenstern theory and Savage development are explained. The contrast between these theories and standard decisions is shown. The irruption of prospect theory is also explored.

Chapter 3 provides an overview of probability theory and discusses the most important concepts used in decision theory.

Chapter 4 debates the origins and foundations of Monte Carlo simulation theory and shows examples of its use in assessing uncertainty.

Chapter 5 elucidates the theory of design of experiments and its use in screening the essential input factors into project valuation, which are later used to drive data acquisition.

Chapter 6 discusses the fuzzy logic theory and contrasts it with Boolean logic. Uncertainty due to fuzziness is addressed and contrasted with uncertainty due to a lack of information. The chapter also includes complete description of the fuzzy inference system.

Chapter 7 is one of the core chapters of the book. It explains in detail the theory of value of information for perfect and imperfect information. Fuzzy data has a dedicated section, and the value of information formalist is adapted to consider the uncertainty due to the fuzzy nature of data, with several illustrative examples shown.

Chapter 8 introduces the concept of the value of flexibility and real options. Several valuations methods are discussed with emphasis on the engineering approach. The fundamental concept of project value improvement through flexibility is explained and examples are shown.

Finally, Chap. 9 discusses four case studies that show, in practical applications, the methodologies introduced in the previous chapters.

Madrid, Spain                                                                                      Martin J. Vilela
Aberdeen, UK                                                                                  Gbenga F. Oluyemi

# Contents

# List of Figures

# List of Tables

# Chapter 1
# Decision-Making: Concepts, Principles, and Uncertainty

**Objective**

This first chapter introduces the reader to the meaning of decision-making, the main concepts associated with decision processes, and the decision analysis workflow. We discuss how to make decisions and the role of uncertainty in the decision-making process. A few valuation methods are discussed. We describe several methods to make rational decisions under certainty, ignorance, risk, and uncertainty. We introduce the concept of vagueness that will be developed in Chap. 6.

## 1.1 Introduction to Decisions

As Hansson (1994) commented, "almost everything that a human being does involves decisions. Therefore, to theorise about decisions is almost the same as to theorise about human activities". This statement summarises how the author believes in the role of decisions in human life.

Of course, when we have a decision problem, we have options to choose from; otherwise, there is nothing to decide.

The founder of decision theory is the French mathematician, physicist, inventor, and Catholic theologian Blaise Pascal (1623–1662). Pascal proposes that decision theory helps us choose the best alternative when there is uncertainty about the future.

Pascal applied decision theory to the problem of the existence of God; he used for the first time the decision matrix shown in Table 1.1.

In this decision matrix, the *States of Nature* are "*God exists*" and "*God does not exist*"; the decision alternatives are "*You bet that He exists*" and "*You bet that He does not exist*"; the consequence is "+∞ *(infinite gain)*", "+x *(finite loss)*", "−∞ *(infinite loss)*" and "+x *(finite gain)*".

Pascal argues that, given this wager, the best alternative is to bet that God exists. If you bet that He exists, you can have infinite benefits (eternal life in paradise) or the contrary, a small loss (avoiding earthly pleasure during your mortal life). However,

M. J. Vilela and G. F. Oluyemi, *Value of Information and Flexibility*, Petroleum Engineering, https://doi.org/10.1007/978-3-030-86989-2_1

**Table 1.1** Pascal's decision matrix for God

|  | You bet that He exists | You bet that He does not exist |
|---|---|---|
| God exists | $+\infty$ (infinite gain) | $-\infty$ (infinite loss) |
| God does not exist | $+x$ (finite loss) | $+x$ (finite gain) |

if you bet that He does not exist, you can have an infinite loss (loss of paradise and the possibility of going to hell) or the contrary, have a limited gain. According to Pascal's reasoning, we have nothing critical to losing by believing even if God does not exist. The probability of God to exist is unknown, but even if you estimate it as exceedingly small ($\varepsilon$), Pascal argues that believing is preferred to no believing.

Surprisingly, for the early envisioning, Pascal's wager introduces for the first time three critical concepts in decision analysis that we discuss later in this book: decision matrix, expected utility maximization, and multiple priors.

Howard (1966) defines a decision as an irrevocable allocation of resources. Irrevocable means that it is impossible or too costly to return to the situation existing before making the decision. An example of an irrevocable resource allocation is drilling a well. Other decisions, such as an investment in public funds, can be revoked but at a cost. This definition of decision pulls apart a decision from a mental proposal to pursue a course of action.

Decisions are taken in many domains: political (which party to vote for), daily life (what clothes to wear, what to have for lunch today, should we go to the cinema today), economic (where to invest our funds, which project to sanction); decisions are also made on the best project to invest in or whether to continue in a venture or not.

In a very naïve approach, when we make decisions, we aim to make *rational decisions*; this statement means two things: we can *weakly order the states between which we can choose*, and we can *make choices to maximise something*.

First, we should say that either we prefer A to B, prefer B to A, or be indifferent between them. Thus, the second requirement for weak order is that all preferences must be transitive: if we prefer A to B and prefer B to C, we prefer A to C. Similarly, if we are indifferent between A and B and between B and C, then we are indifferent between A and C.

The second requirement for a rational decision-maker is that he must choose to maximise something; in the theory of risky choices, the object to maximise is the expected utility. This maximisation means that the decision-maker selects the best alternative or set of alternatives based on some attribute that he cares about. Of course, maximisation can be reached in several forms, and it is crucial to select the appropriate maximisation approach.

Decision-making is done using entirely different approaches, from the most informal ones such as intuition to formal ones such as the expected utility value. Some of these are discussed in this chapter and Chap. 2 of this book.

Decision-making is a fundamental part of any process or venture, economic, scientific, political, or social. Decision-making aims to make consistent and correct choices that follow the decision-maker's criteria in an uncertain environment. Because making a decision is a critical activity in the Exploration and Production (E&P) oil business, geoscientists and petroleum engineers must produce information to support decisions (Bratvold et al., 2007).

Howard (1968) described decision analysis as the formal introduction of logic and preferences into decisions through a combination of philosophy, methods, practice, and applications. Decision-making is an important management activity, and, very often, it is a difficult task (Taghavifard et al., 2009).

When we make decisions in a certain environment, the best decision always yields the best outcome. However, as we will discuss later, that is not always the result in an uncertain environment.

When making a decision, we should distinguish between *a good decision* and *a good outcome*. When we make a rational and logically consistent decision with the available information, uncertainties, decision-makers' values, beliefs, and preferences, we are making a good decision (Howard, 1966). A good outcome is a future state of the world that we prefer over other possible ones; however, even if the decision is good, the outcome can be bad. Indeed, the situation can also be the opposite: bad decision-making can lead to good outcomes. Making good decisions does not secure good outcomes in a specific case, but it will do so in the long run: more "goods" than "bads" will happen if we make good decisions or ensure a high percentage of good outcomes. As described by Bratvold et al. (2002), the best hope for a good decision outcome is a good decision process. Therefore, the goal of decision analysis is to focus on making good decisions, which, in the long run, should result in an increased number of good outcomes. Another characteristic of the decision process is that most of the time, the decision is made under incomplete or uncertain information, making the outcomes uncertain.

Begg et al. (2003) argued that uncertainty is the reason behind the failure of many projects to achieve their optimal performance; uncertainty leads to over-estimating benefits or under-estimating the chance of losses; in the same research, these authors mention Marrow's study of more than 1,000 E&P projects with capital expenditures from US$1 million to US$3 billion (average US$679 million), which found that many failed to produce the expected results and 12.5% were a "financial disaster" because they did not meet 2 out of 3 metric items: (i) cost greater than 40% of the plan, (ii) time slippage greater than 40%, and (iii) the operability of the project after one year was less than 50% of the plan.

As we discuss in Chap. 2 of this book, most people are risk-averse. This means that when the decision involves the possibility of risking a large proportion of their wealth, injury or another form of fundamental thread, people do not follow rational decisions, as described by von Neumann and Morgenstern's theory, but develop many biases as described in the Prospect theory, discussed in Chap. 2 of this book.

## 1.2 Overview of Decision Analysis

Decision analysis is a method that balances the factors impacting a decision (Howard, 1966). Bickel and Bratvold (2007) describe decision analysis as a systematic method that transforms untransparent decision problems into transparent ones by applying a sequence of defined steps; decision analysis helps decision-makers select the optimum course of action for the concerned problem.

Cunha (2007) defines decision analysis as a structured way of thinking for visualising the possible alternatives available to solve a problem; in the decision analysis frame, the decision-maker assesses the consequences of the possible alternatives and their corresponding likelihood.

Decision analysis aims to provide decision-makers with methods for making better decisions in complex situations. The decision analysis process steers decision-makers to think systematically about their objectives, preferences, and the problem's uncertainty and structure, helping to model it qualitatively.

Decision analysis techniques have been applied extensively to many assessments in the oil and gas industry, such as in production performance evaluation (Vanegas et al., 2005), field development constrained to geomechanics aspects of the reservoir (Rodrigues et al., 2006), well cost estimation (Cunha et al., 2005), and waterflood risk assessment project (Peake et al., 2005) among others.

All the activities that a human being undertakes involve making decisions. Decision theory studies the process of making decisions and analyses the decision-maker's criteria. Decision theory can be *normative, prescriptive,* or *descriptive.*

*Normative decision theory* studies which decisions must be taken by a rational and coherent individual; it indicates how decisions should be made (rational decision-makers). *Normative* rules provide the framework to deal with complex decision problems where the limitations of our cognitive abilities can mislead us. *Descriptive decision theory* studies how individuals make decisions in their everyday lives; descriptive theory builds theories that explain people's behaviours, attitudes, and processes when making decisions. It provides experimental evidence that people often make decisions in ways that are inconsistent with rational thinking. *Prescriptive decision theory* studies how real individuals make decisions, given their cognitive limitations and lack of information. *Prescriptive* also means that it suggests how decisions should be made in practice, given that decision-makers are not entirely rational individuals. *Prescriptive* decision theory is between *normative* and *descriptive* theories, keeping solid roots in the *normative* process.

Authors (Hansson, 1994) who consider decisions, as far as subjects of decision theory, should include only *normative* decisions, and everything outside what is a rational decision should be excluded from decision theory (including ethical and political norms).

Decision theory provides the method to achieve the goal, but it does not question the reason for the goal.

To distinguish between the *normative* and *prescriptive* approaches, Helbert Simon said that the *normative* approach is what the decision-maker would do without all constraints, such as time, information, etc. In contrast, the *prescriptive* approach describes a feasible procedure for the decision-maker to follow.

Several authors, notably Condorcet (1743–1794), John Dewey (1859–1952), Herbert Simon (1916–2001) and Brim (1923–2016), propose a modified sequence or steps for decision-making as an attempt to systematise the decision process. These theories consider decisions as a sequential process. Mintzberg et al. (1976) develop a parallel approach (non-sequential) to make decisions that better represent the actual decision process (see references for more detail).

In a decision problem, we have a decision-maker who has a set of preferences; those preferences should satisfy criteria of logical consistency:

(1)  *Transitivity*: for all x, y and z, if x is strictly preferred to y and y is strictly preferred to z, then x is strictly preferred to z.
(2)  *Completeness*: for all x and all y, either x is preferred to y or y is preferred to x or the decision-maker is indifferent between them.
(3)  *Asymmetry*: if x is strictly preferred to y, y is not strictly preferred to x.
(4)  *The symmetry of the indifference*: for all x and y, if x is indifferent to y, y is indifferent to x.

If the decision-maker follows these criteria when making decisions, we say that he chose rationally or logically. Therefore, he has a utility function, i.e., a number associated with each preference, to be ordered from less preferred to more preferred.

For making decisions, the decision-maker has a set of *States of Nature* or scenarios, a collection of alternatives and a set of consequences. The *States of Nature* are the possible outcomes in a decision situation over which the decision-maker has no control. The alternatives are the possible course of action that the decision-maker can take, and these are in his control. The consequences or payoffs are the "values" associated with any combination of alternatives and the *State of Nature*. They can be anticipated and ordered according to the decision-maker's preferences. A rational decision-maker will choose the alternative with the best consequence.

Alternatives are courses of action that are possible for the decision-maker. When the decision-maker knows the consequences of the several alternatives, the information is complete. Thus, the decision-maker is in a situation of certainty; otherwise, when the information is incomplete, the situation is uncertain.

For making decisions, several criteria have been used for choosing between alternatives; for a criterion to be well-defined, it should be able to prescribe a precise algorithm that selects, without ambiguity, the optimal alternative. To make decisions, we assume that we have a set of acts or alternatives $A_1, A_2, \ldots, A_m$, a set of states $s_1, s_2, \ldots, s_n$ and the corresponding utility values $u_{ij}$, $i = 1, \ldots, m$ and $j = 1, \ldots, n$. The decision criteria are the basis for making the decision, and we make the best choice (or decide) by weighing each alternative against the criteria established.

The objective of a decision problem is to produce an optimal quantity of variables that the decision-maker cares about; in many problems, there are two or more

**Table 1.2** Decision matrix
for the umbrella example

| | It rains | It does not rain |
|---|---|---|
| Take an umbrella | Not wet, more weight | Not wet, more weight |
| Do not take an umbrella | Wet, less weight | Not wet, less weight |

objectives which, in some cases, can conflict: an alternative that is optimum for one objective may be suboptimal for other alternatives; in these cases, the decision-maker should define, upfront, his preferences based on trade-offs amongst the alternatives.

In this book, we study decisions between several options but taken in isolation by individuals. Game theory studies a broader problem: individuals making decisions influenced by the decisions of other individuals or strategic decisions. These problems will not be considered in this book (for the Game theory problem, refer to the reference von Neumann and Morgenstern, 1944).

In making decisions, the simplest case is when the results are certain, which means that each alternative has only one possible outcome. For this case, the rational decision is to select the alternative with the highest value. This approach is called the Maximum Return Criteria. However, assessments become more complicated when uncertainty and risk are present.

**Example 1.1: Decision: Rain**  Assume that I plan to leave my house and two alternative weather conditions may occur, it may *rain* or *no rain,* and these are the two *States of Nature* in this problem. The *alternatives* are the decisions available to me, which in this example are: *take an umbrella* or *do not take an umbrella*. If I decide to take the umbrella with me, I will carry more weight. The consequences are the effects that my decision brings, which depend on the states and alternatives chosen.

Decisions can be represented by the matrix shown in Table 1.2.

Consequences can also be represented by numbers that capture their desirability.

## 1.3  Uncertainty, Risk, Vagueness

Only a few decisions are taken where there are no uncertainties in some or most of the input parameters used for making decisions. The decision analyst must assess these uncertainties' magnitude, impact, and consequences (Begg et al. 2003).

As Mark Dean (Columbia University) mentioned, a roulette wheel is an example of risk, and a horse race is an example of uncertainty. In a roulette wheel, there are 38 slots; someone who places a US$10 bet on number 7 is in a lottery which pays US$350 with a probability of 1/38 and zeroes otherwise (US$30 goes to the house in each trial). In a horse race with three horses, someone who places a US$10 bet on a horse does not necessarily have a 1/3 chance of winning.

Most people make distinctions between situations like tossing a coin or throwing dice governed by objective probabilities and problems like the likelihood of a nuclear war or the price of copper in 20 years that do not have associated objective probabilities.

Uncertainty is a characteristic of a variable that is not known with certainty; when a variable is uncertain, it is impossible to know its *State of Nature* or a future outcome precisely.

Some authors use the terms risk and uncertainty interchangeably, and others use risk to refer to an adverse event (the risk of losing US$1,000,000).

However, there is a clear distinction between these concepts; the American economist Frank Knight (1885–1972), one of the founders of the famous Chicago School of Economics, formalised in his book *Risk, Uncertainty, and Profit* (1921) the distinction between uncertainty and risk:

(i)   the risk applies to situations where, for a given decision, the alternatives are known, the outcomes of the decision are unknown, but the odds can accurately be measured, and the probability distributions are known,

(ii)  uncertainty applies to situations where the alternatives of a decision are known, the outcomes of the decision are unknown. Because of lack of information, the odds and probability models are also unknown.

For Knight, the distinction between risk and uncertainty is critical for understanding the behaviour of the financial world. Knight defines "relevant" probabilities, those used in risk situations, as those obtained by (1) a *probability* a priori based on mathematical calculus—based on equally probable phenomena, used in theoretical situations but challenging to implement in business cases—and (2) a *statistical probability* based on experience—empirical and derived from real facts, based on past and similar real case situations. This approach assumes that what happened in the past holds in the future, which is an assumption that needs to be verified in each case.

According to Knight, there is a third kind of probability, a *probability estimation* based on empirical data unrelated to mathematical models or previous statistics. This third form of probability is the subjective probability, very frequently used in decision problems but not described in detail by Knight. Instead, he used the terms "measurable uncertainty" and "unmeasurable uncertainty" to refer to what we call risk and uncertainty and the objective and subjective probability of the corresponding probabilities.

To summarise, the practical difference between risk and uncertainty is that, in the former, the distribution of the events in the problem is known (either with an a priori calculation or based on statistics of previous cases) but, for the latter, it is not possible to compute or assign probabilities with certainty.

Even though the distinction between risk and uncertainty is very relevant, it has no meaning within the framework of decision theory, i.e., in Savage's theory of decision (Chap. 2). This happens because, in decision theory, the decision-maker reduces all uncertainties to risks by using the expected utility criterion with respect to his subjective probabilities. In Savage's theory, probability is purely subjective.

Uncertainty can be due to the stochastic nature of the variable (no single value exists), incomplete knowledge (ignorance of parts of the complete system), error in data (inaccuracy in the data measured), conceptual imprecision (lack of clarity on what the measure is), lexical vagueness and linguistic inexactitude.

Another form of uncertainty is vagueness. Vagueness was discussed by the British mathematician and philosopher Bertrand Russell (1872–1970) by using the paradox of a bald person: initially, a man has a full head of hair, but he gradually loses his hair until he is bald; in between these two extreme cases, when can we say that the man is bald? Is it when he has no hair, or can we consider him bald when he has just ten hairs left? This paradox shows an example of vagueness. There are men who we can say, without doubt, are bald, and others, still with most of their hair, who we can undoubtedly say are not bald, but the concept of baldness is vague between the extremes.

The vagueness in the variables can be represented by using the concept of partial membership. As we will discuss in Chap. 8 of this book, membership is a mapping between the set of variables and the interval [0, 1], where 0 means "no membership", 1 means "full membership", and any number in between those means "partial membership".

Vagueness is the result of the way humans think and, as such, is related to linguistics or our natural language. We think in categories, but a number of these, discrete or continuous, describe objects in the external world. The authentic paradox between the way the external world is represented and our mental projection of reality creates vagueness.

Bellmann and Zadeh (1970) state a clear distinction between uncertainty due to randomness, which can be dealt with using probability theory and fuzziness, which use fuzzy theory. For example, the class of "tall men" is a fuzzy set because this set is qualified with an impression; in that sense, an individual who is 180 cm tall has a degree of membership in the set of tall men, and that degree of membership is measured with a real number between 0 and 1. Fuzzy sets use adjectives such as simple, approximate, significant, etc., characterised by not having sharp boundaries. Fuzziness is related to classes of objects characterised by membership between full membership and no membership. On the other hand, randomness handles uncertainties concerning an object's belonging or non-belonging to a set. For example, we can say that an individual has a 40% chance of being taller than 180 cm. In this statement, we divided men into two groups, taller than 180 cm and shorter than 180 cm, and we state that this individual has a 40% chance of belonging entirely to the set of men taller than 180 cm.

Uncertainty, in any of its forms, is present in many problems. Demirmen (2001) correctly said that reservoir uncertainty is inherent to all hydrocarbon prospects and fields. Indeed, technical teams spend a significant amount of time and effort to understand it and derive the consequences and impacts that uncertainty has on developing the assets. Demirmen's focus is the reservoir uncertainties at the appraisal phase of a field, where uncertainties are classified based on (i) the volume of hydrocarbon in place such as the reservoir structure and petrophysical properties, (ii) recovery factor/productivity such as residual oil saturation and aquifer support, and (iii) fluid properties such as gas composition and crude viscosity.

In the survey conducted by Bickel and Bratvold (2007) amongst oil and gas professionals (completed by 494 professionals during April–July 2007), it is shown that the primary sources of uncertainties having a potential impact on the performance of E&P projects are subsurface risks, hydrocarbon prices, and reserve estimates.

Some authors (Virine & Rapley, 2003) believe that, even though decision theory has been accepted in the petroleum industry, its application in real-world problems is limited because of the complexity of the tools available for making decision analysis.

## 1.4   Review of Methods for Decision-Making

In the framework of decision analysis, each decision has three components:

(A)   *A set of alternative actions*: these are the options available to the decision-maker.

(B)   *A set of possible States of Nature*: these are the states of the system that are possible; only one will be the outcome of the trial, but we do not know which one.

(C)   *A set of outcomes*: one for each combination of an alternative and a *State of Nature*.

The decision-maker chooses alternatives that have outcomes; these outcomes depend on the alternatives and *States of Nature*.

In the literature, the decision-making process is described by a series of steps (Begg et al. 2003):

(i)    Define the decision problem.
(ii)   Select the alternatives.
(iii)  Define the criteria that will be used to assess the alternatives.
(iv)   Establish a weighting factor for the criteria defined in step iii.
(v)    Assess each alternative.
(vi)   Estimate the value of each alternative and select the one with the highest value.
(vii)  Perform sensitivity analysis.

In some cases, steps (i) and (ii) or (v) and (vi) are merged; step vii is not always done.

After selecting one alternative, the true *State of Nature* reveals itself. The decision-maker obtains the corresponding outcome (the one associated with the chosen alternative and the *State of Nature* revealed). Thus, decision analysis is a technique that helps to choose consistently between alternatives according to given criteria.

Luce and Raiffa (1957), two of the most fundamental researchers in decision theory, describe three different situations for decision-making:

(1)  Certainty: decisions under certainty mean that each alternative is known to lead to a specific outcome. In decision-making under certainty (DMUC), the decision-maker knows with certainty what the *State of Nature* will be, and he knows it in time to make the best choice. In a certainty environment, if the decision-maker chooses alternative A, outcome Y will occur, and if alternative B is selected, outcome Z will occur.

(2)  Risk: decision under risk means that each alternative leads to one of a set of possible specific outcomes, each one with a known probability, which is assumed to be known by the decision-maker. In decision-making under risk (DMUR), the decision-maker does not know what the *State of Nature* will occur but can assign an objective probability to each *State of Nature*.

(3)  Uncertainty: in the decision under uncertainty, the decision-maker knows the *State of Nature,* and each alternative leads to one element in the set of possible outcomes, each with a subjective probability estimated based on the beliefs, experience, and degree of knowledge of the decision-maker. In decision-making under uncertainty (DMUU), the decision-maker does not know what the *State of Nature* will occur but can assign a subjective probability to each *State of Nature*.

In a decision under uncertainty, the decision-maker does not know the outcome when an alternative is selected. For example, if a decision-maker selects alternative A in an uncertain environment, two or more outcomes may occur. If alternative B is chosen, again, two or more outcomes are possible.

Luce and Raiffa acknowledge that most decisions fall between risk and uncertainty. We may have partial knowledge or experience that leads us to estimate, without certainty, what the probability is for the different outcomes. In Example 1.1 about the weather conditions, we may assign the probability of rain based on a weather forecast or intuition.

Due to this, the fourth category of decisions is

(4)  Ignorance: decision under ignorance means that each action leads to any element in the set of possible specific outcomes with a completely unknown probability. In decision-making under ignorance (DMUI), the decision-maker knows the *State of Nature*, but he does not know any estimate of how probable a state will be.

Wu et al. (2004) propose that decisions under risk occur when probabilities are objectives, such as those resulting from tossing a fair coin or rolling a balanced die, while decisions under uncertainty occur when the chances are subjective, such as investment decisions or decisions depending on the future forecast.

Decision-making under certainty

There are two cases in which making a decision is easy:

(i)  Under certainty: if we know what the *State of Nature* will be, we only need to compare each alternative which outcome, associated with the sure *State of Nature*, produces the more significant outcome.

(ii)     Under stochastic dominance: if alternative A has a better outcome than alternative B under each individual *State of Nature*, then alternative A is stochastically dominant over alternative B. Assuming that what the decision-maker wants is to get the best outcome, he will always choose alternative A instead of alternative B.

The selection of the optimum alternative is made based on a decision rule. A decision rule is a mapping from the outcomes to the set of real numbers representing a value or a utility value; each outcome depends on the corresponding alternative and *State of Nature*.

In cases where there is a certainty, making a decision is, in general, straightforward. First, we need to define the objective function, which is the one that establishes the value or what is essential for the decision-maker. Most often, people are concerned with money, which is typically the objective function. Second, we should have a forecast of the benefits and costs during the period of the assessment; the benefits such as the oil production or the number of cars produced during a fixed time period are translated into money; similarly, a forecast of the future investments and operating cost associated with the benefits forecast should be estimated. Third, the cash flow method accounts for the yearly net cash, and the net present value (NPV) converts the future cash flow into money today using the discount rate factor. Finally, the NPV or other metrics such as the internal rate of return (IRR) assess the value of the project. Because there is no uncertainty in the input or the output, the value of the financial indicator will tell whether it is worthwhile to move the project forward.

The net present value (NPV) is the present value of the expected future cash flows, discounted at the selected discount rate, subtracting the initial cost of the investment. The discount rate is used to convert the future values to present ones.

$$NPV = \sum_{j=1}^{n} \frac{values}{(1+i)^j} \tag{1.1}$$

where $i$ = discount rate, *values* are the yearly result of adding benefits and costs, $j$ is each period (year), and $n$ is the total number of years.

In most cases, cash flows are estimated yearly; the initial investment is typically included in the first year, and for subsequent years benefits and costs are added.

As an example, assume that a project requires an investment of US\$3,000,000, and we are sure that with such investment, we can produce, during the next five years, the following benefits: US\$2,000,000, 3,500,000, 3,400,000, 1,700,000 and 1,200,000. Then, using a discount rate of 10%, the value of this project is US\$6,171,443, which is a positive value and indicates that it is worthwhile to proceed with this project. However, if the initial investment is US\$10,000,000 instead of US\$3,000,000, the value of the project is US\$−828,557, which is a negative value indicating that the project will produce losses.

Another popular indicator is the IRR, which estimates the discount rate required to make the NPV of the project equal to zero, indicating the attractiveness of a project.

When a project's input parameters have uncertainties, the output parameters will inevitably also carry uncertainties.

The input variables can have discrete or continuous values producing continuous or discrete outcomes; in any case, it is crucial to use a small number of discrete cases to solve practical problems. For this reason, it is pretty common to consider three cases: low, medium, and high, which in other contexts, people refer to as pessimistic, mean, and optimistic. For example, suppose a project depends on several input parameters, each one with uncertainty. In that case, the objective function is discretised as low, medium, and high when the combined effect of the input variables produces small, medium, and high values. This approach is subjective.

### Decision-making under ignorance

For decision-making under ignorance, we know the *States of Nature* but not their probabilities; several decision criteria are independent of the probability of the outcomes, and these approaches depend on the values of the outcomes.

### Maximin criterion

The Maximin criterion assigns each alternative its minimum value then selects the alternative with the greatest minimum value. Thus, each alternative is assessed to the worst state of that alternative, and the optimum choice is the one with the best worst state. This criterion ensures that if the worst occurs, you have selected the best alternative. It emphasises *States of Nature* that are unfavourable. Thus, the Maximin criterion is a conservative and pessimistic criterion. We can call it the fear criterion.

The application of this criterion can lead to a contradiction with common sense.

**Example 1.2: Maximin Criterion: Basic**  In this example, the decision-maker has two alternatives $A_1$ and $A_2$, two *States of Nature* $S_1$ and $S_2$ and the consequences, as shown in Table 1.3.

When using the Maximin criterion, 0 is assigned to alternative 1, 1 is assigned to alternative 2, and the final assessment selects the alternative 2, which has the maximum of the worst.

The drawback of this criterion can be observed in that alternative 1 has the possible outcome of 100,000, which is much higher than the rest, and that value is discarded; also, the utility value of $s_1$ in alternative 2 can be replaced by 0.0001, leading to the same selection of alternative 2.

The Maximin criterion chooses the option that has less to lose.

**Example 1.3: Maximin Criterion: Pessimism**  An investor has $10,000 and two possible investment options: (i) invest the resources in a savings bank that pays a sure $1,400 after one year (14% interest rate), and (ii) invest in the stock market.

**Table 1.3**  Example 1.2: Maximin criterion: basic

|       | $S_1$ | $S_2$   |
|-------|-------|---------|
| $A_1$ | 0     | 100,000 |
| $A_2$ | 1     | 1       |

**Table 1.4**  Example 1.3: Maximin criterion: pessimism

| Action | State of Nature | | | Maximin decision |
|---|---|---|---|---|
| investment options | Market up | Market stable | Market down | |
| Savings account | 1,400 | 1,400 | 1,400 | 1,400 |
| Funds | 3,600 | 500 | −500 | −500 |
| | | | | **1,400** |

For this analysis, we will consider three possible scenarios for the stock market's performance: market up, market stable and market down, corresponding to the cases where the return of the stock market is large, medium, or low. Table 1.4 shows the possible outcomes for this investment and the decision using the Maximin criterion.

The Maximin criterion decides on the highest benefit, which is assured. In this example, the savings account secures the same benefits independently of the fluctuations in the market. However, this approach means that the decision-maker loses the opportunity that originates from the possibility that the market moves up, in which case he would gain $2,600. This loss of opportunity is the price to pay for having a secure return; that is why many people have called this approach a pessimistic one.

Minimax regret criterion

The Minimax regret criterion selects the alternative that will minimise the maximum regret; regret means the opportunity loss from selecting an alternative other than the one that turns out to be the best. Opportunity loss is the difference between the actual value for a decision and the optimal value for the same *State of Nature.*

For applying this criterion, the utility values of each state are replaced by the amount that has to be added to equal the maximum utility value for that state. This procedure is repeated for each state until it generates the regret or opportunity-loss matrix; to each alternative, we assign its maximum regret. Then the alternative selected is the one with the minimum of the maximum regrets.

The Minimax regret criterion strives to emphasise the *State of Nature* where our actions make the most difference by selecting the alternative with the minimum opportunity loss (or regret).

**Example 1.4: Minimax Regret Criterion: Basic** To use the Minimax regret criterion, we should build the regret matrix, shown in Table 1.5.

In Example 1.4, on the left, we have the utility payoffs, and on the right, the risk payoffs, regret matrix or losses opportunity table. The criterion assigns to the

**Table 1.5**  Example 1.4: Minimax regret criterion: basic

| | S1 | S2 | | | S1 | S2 |
|---|---|---|---|---|---|---|
| $A_1$ | 0 | 100,000 | ----➔ | $A_1$ | 1 | 0 |
| $A_2$ | 1 | 1 | | $A_2$ | 0 | 99,999 |

**Table 1.6**  Example 1.5: Minimax regret criterion

| Action | State of Nature | | | Minimax regret decision |
|---|---|---|---|---|
| Investment options | Market up | Market stable | Market down | |
| Savings account | 2,200 | 0 | 0 | 2,200 |
| Funds | 0 | 900 | 1,900 | 1,900 |
| | | | | **1,900** |

alternative A1 the index 1 and the alternative A2 the index 99,999; the criterion selects alternative A1 as the minimum regret index. By using this criterion, the decision-maker is sure to minimise the losses.

**Example 1.5: Minimax Regret Criterion**  In this example, we will use the same data discussed in Example 1.3, but we will use the Minimax regret criterion. The first step is to build the losses opportunity table, as shown in Table 1.6.

The action with less regret is to invest the money in funds.

The basis for making decisions using this criterion is the minimisation of opportunity losses.

Maximax criterion

The Maximax criterion selects for each alternative the maximum value and then chooses the alternative with the greatest maximum value. Thus, each alternative is assessed to the better state of that alternative, and the optimum choice is the one with the best of the better states. This criterion selects the alternative which, if the results turn out to be the best, provides the optimum value; it emphasises the favourable States of Nature. This criterion is optimistic because it selects the alternative based on the assumption that the result will be the highest possible.

**Example 1.6: Maximax Criterion**  We will use the data in Example 1.2 and the Maximax criterion, which is applied as shown in Table 1.7.

The Maximin and Minimax regret criteria are pessimistic (assessed based on the worst alternatives), while the Maximax criterion is optimistic (estimated based on the best alternatives).

The pessimism–optimism criterion of Hurwicz

The Hurwicz criterion (1951, Leonid Hurwicz, 1917–2008, Polish-American economist and mathematician) provides a formula combining the best and worst

**Table 1.7**  Example 1.6: Maximax criterion

| Action | State of Nature | | | Maximax decision |
|---|---|---|---|---|
| Investment options | Market up | Market stable | Market down | |
| Savings account | 1,400 | 1,400 | 1,400 | 1,400 |
| Funds | 3,600 | 500 | −500 | 3,600 |
| | | | | **3,600** |

**Table 1.8**  Example 1.7: Hurwicz criterion, $\alpha = 0.7$

| Action | State of Nature | | | Hurwicz decision |
|---|---|---|---|---|
| Investment options | Market up | Market stable | Market down | |
| Savings account | 1,400 | 1,400 | 1,400 | 1,400 |
| Funds | 3,600 | 500 | −500 | 2,370 |
| | | | | **2,370** |

alternatives. It uses the weighted average of the worst and the best outcome per alternative; then, the selected alternative is the maximum value.

The weight is defined as the $\alpha$ index, an optimistic coefficient, which is a number between 0 and 1, where $\alpha$ is the weight used for the best outcome and $1 - \alpha$ is the weight used for the worst outcome of each alternative. This weight factor is known as the pessimism–optimism index; when it is equal to 1, the Hurwicz criterion is the same as Maximax, and if equal to 0, it is the same as Maximin.

**Example 1.7: Hurwicz Criterion**  The result of using the Hurwicz criterion in the data discussed in Example 1.2 is shown in Table 1.8.

The "principle of insufficient reason" of Laplace or the Laplace–Bayes rule

The principle of insufficient reason is also called the Principle of Equal Likelihood. Suppose that you are entirely ignorant of each alternative and unable to give the associated weight (likelihood) for the states according to this criterion. In that case, you should assign the same weight (equally likely) to each state. The principle of insufficient reason treats all *States of Nature* as being equally likely to happen. The index of each alternative $A_i$ is,

$$\frac{u_{i1} + u_{i2} + \cdots + u_{in}}{n}$$

The best alternatives are the ones with the highest values.

**Example 1.8: Principle of Insufficient Reason Criterion**  In this example, we use the data from Example 1.2, but the decision is assessed using the principle of insufficient reason, as shown in Table 1.9.

Neither the Minimax nor Maximin criteria allow the decision-maker to use his personal opinion, experience or beliefs about the likelihood of any of the alternatives

**Table 1.9**  Example 1.8: principle of insufficient reason

| Action | State of Nature | | | Principle of insufficient reason decision |
|---|---|---|---|---|
| Investment options | Market up | Market stable | Market down | |
| Savings account | 1,400 | 1,400 | 1,400 | 1,400 |
| Funds | 3,600 | 500 | −500 | 1,200 |
| | | | | **1,400** |

available occurring. However, in most real-world cases, the decision-maker has some knowledge that can be used as a guide in the decision-making process, and this fact is not included at all with any of these four criteria discussed.

## Swanson's 30-40-30 rule

Roy Swanson published an Exxon oil Company communication (1972) proposing to use the 30-40-30 rules for assessing the average estimate of reserve for exploration prospects; Hurst et al. (2000) published this proposal.

In exploration evaluation, people often use the triple point approach: the *most likely reserve estimate* in a prospect is estimated based on the product of the midpoints of three factors: areal extent, pay thickness, and recovery factor. Typically, after this *most likely estimate*, the ranges of the reserves are obtained using Monte Carlo simulation (Chap. 4) on those factors.

What Swanson proposed is to use a prescribed procedure to estimate the average value based on the following rule: the average value of the reserve for a prospect is obtained as the expected value of three *States of Nature*: $P_{10}$, $P_{50}$ and $P_{90}$ weighted with the numbers 0.30, 0.4 and 0.30, respectively. The main aim of the proposal is to use the average value as prescribed instead of using the most expected value based on the medium of the input parameters.

Swanson recognised that people use different terminology for $P_{10}$, $P_{50}$ and $P_{90}$ but he used the following one:

(i)     $P_{10}$ (minimum) to indicate 90th percentile (90% of cases are larger than $P_{10}$)
(ii)    $P_{50}$ (medium) to indicate 50th percentile (50% of cases are above, and 50% of cases are below $P_{50}$)
(iii)   $P_{90}$ (maximum) to indicate the 10th percentile (10% of cases are more significant than $P_{90}$)

The rationale for the 30-40-30 rule is: for $P_{50}$ the ranges to $P_{10}$ and $P_{90}$ are 0.40 each (the full range is 1.00 and a linear split between the medium value and the maximum and minimum values (0.40 as defined). Swanson assumed that $P_{50}$ takes half of these two intervals, i.e., 0.20 below and 0.20 above (indeed, this is the primary assumption for getting the rule); the remaining 0.20 on each side is taken for $P_{10}$ and $P_{90}$; finally, each of these extreme values has 0.10 on each side. Following this rationale, the rule assigns: 0.30 probability to $P_{10}$, 0.40 probability to $P_{50}$ and 0.30 probability to $P_{90}$.

Under the assumption supported by experience and real cases discussed by Hurst, Brown and Swanson, reserves in oil exploration prospecting can be described with a Lognormal distribution. They propose a graphical method to estimate the $P_{10,}$ $P_{50}$, and $P_{90}$. In a Lognormal distribution, taking the midpoint to represent the most likely reserve is not correct. For that reason, Swanson proposes to use the average prospect size rather than a prospective most likely case.

The value of $P_{10}$ is calculated as the minimum reserve held by the one-well project, which recoups the cost of this venture. The $P_{50}$ value is estimated to be equal to the most likely case (which in general is not true but an approximation). Assuming that

**Table 1.10** Example 1.9: Swanson rule criterion

| Action | State of Nature | | | Maximax decision |
|---|---|---|---|---|
| Investment options | Market up | Market stable | Market down | |
| Savings account | 1,400 | 1,400 | 1,400 | 1,400 |
| Funds | 3,600 | 500 | −500 | 1,130 |
| | | | | **1,400** |

the reserves satisfy a Lognormal distribution and plotting on log plot probability versus reserves, the value of $P_{90}$ can be graphically calculated. If the value for $P_{90}$ is too large or too small based on "reasonable" values for the input parameters, $P_{50}$ can be fine-tuned to accommodate a more feasible $P_{90}$. However, the procedure just described is not strictly related to Swanson's rule and is more a method for the explorations analyst to estimate reserves without using other methods to estimate the range of reserve variability, such as Monte Carlo simulation.

**Example 1.9: Swanson Rule Criterion** Even though Swanson's rule is not designed for this kind of problem, we use it here to show the evaluation methodology using this rule. Table 1.10 shows the result of applying this criterion for making decisions.

Decision-making under risk and uncertainty

In situations of decision under risk and uncertainty, the decision-maker knows the likelihood that the *States of Nature* will occur; this information is revealed by historical relative frequency information by analysing the phenomena observed or may also be derived subjectively.

Much of the theory of decision analysis is founded on two subjective concepts: our feelings about the uncertainties of the *States of Nature*, assigned by probabilities, and the utility, a measure of the desirability of an outcome. Subjective probability is a function of our information at any given time. The utility is a measure of our preferences which, of course, depend on our situation (wealth, health, etc.) at the time of the decision. For making a decision under risk and uncertainty, the most used criterion is the expected value.

Suppose a decision has two possible outcomes, $x_1$ and $x_2$ with probabilities of occurrence $p_1$ and $p_2$ respectively; the expected value $EV$ is given by Eq. (1.3),

$$EV = x_1 p_1 + x_2 p_2 \qquad (1.3)$$

For economic metrics, a shared decision criterion is the Expected Monetary Value of alternative $j$ defined as

$$EMV_j = \sum_{i=1}^{n} P(S_i) V_{ij} \qquad (1.4)$$

where $EMV_j$ is the expected monetary value of alternative $j$, $P(S_i)$ is the probability of *State of Nature $S_i$* and $V_{ij}$ is the value of the *State of Nature $S_i$* for the alternative $j$. According to this decision criterion, the decision-maker should select the alternative with the highest expected monetary value.

The expected monetary value is a long-term criterion: if we apply it consistently, we will be better than if we do not use it. The risk associated with a particular alternative is its variance:

$$Variance_j = \sum_{i=1}^{n} P(S_i)\left(EMV_j - V_{ij}\right)^2 \tag{1.5}$$

Different decision-makers have a different appetite for risk. The larger the variance of an alternative, the higher the risk is. Therefore, most of the time, the decision-maker prefers to have the EMV as significant as possible and the variance as small as possible.

Using EMV as a decision criterion assumes several conditions:

(A)   The expected monetary value is an average value over many trials; it does not refer to the outcome of one particular trial.

(B)   This criterion does not consider risk: during the trials, some outcomes can be too positive or negative, but whatever the results, the decision-maker continues with the trials.

(C)   The expected monetary value assumes that the decision-maker is risk-neutral, which means a linear function of money can describe his risk attitude: an increase in gain from \$1 to \$10 is equally desirable as an increase from \$1,000,000 to \$1,000,010.

(D)   The expected monetary value measures just one criterion to select the best alternative.

(E)   This criterion considers that the full range of potential outcomes is known and covered by the analysis.

For deciding under uncertainty, the aim is to optimise the benefits yielded by the project, system or process. Two complementary approaches for optimising the project value are discussed in this book: increasing the value by acquiring data related to the project, i.e., the value of information, and increasing the value by creating flexibility in the project, i.e., the value of flexibility.

Several concepts, methods and techniques are fundamental to understand these approaches: expected value, utility functions, probability theory, Monte Carlo simulation, fuzzy theory, design of experiments. We will discuss all of them and their relationship and application to decision theory. In the last three chapters, we will integrate them all by applying these techniques to decision problems under uncertainty.

Our focus is on decision problems related to the oil and gas industry. However, the approach can be naturally extended to any other domain.

## 1.5 Summary

This chapter deals with the decision-making process and sets the methodologies for decision-making. The chapter starts with a historical review of decision theory and explains its importance. Due to its relevance for an effective decision-making process, a section is dedicated to describing the foundations of the discipline called decision analysis; we discuss the meaning of decision-making and the requirements for a rational decision. It clarifies that the practitioners need to understand the distinctions between good decisions, good outcomes, and their relationships. The following section elaborates on the crucial differences between normative, descriptive, and prescriptive decision theories and the scope of their applications. The principal terms used in its framework to effectively apply decision analysis theory are explained, and several examples are presented. A section is dedicated to exploring the concepts of uncertainty, risk and vagueness and their differences. An essential section of this chapter is dedicated to discussing the workflow of the decision-making process and its logic. Finally, it is acknowledged that, depending on the knowledge available at the time of making a decision, several possible decision situations with their corresponding decision methodologies are available, which are discussed in this chapter: decision-making under certainty, decision-making under ignorance and decision-making under uncertainty; several examples are used to explain their differences.

## References

Begg, S., Bratvold, R., & Campbell, J. (2003). Shrinks or quants: Who will improve decision-making. *Society of Petroleum Engineers*. In *Proceeding of the SPE Annual Technical Conference and Exhibition*, Denver, Colorado, USA, 5–8 October 2003. Paper SPE-84238.

Bellmann, R., & Zadeh, L. (1970). Decision-making in a fuzzy environment. *Management Science, 17*(4), B-141–B-164.

Bickel, J., & Bratvold, R. (2007). Decision making in the oil and gas industry: From blissful ignorance to uncertainty-induced confusion. *Society of Petroleum Engineers*. In *Proceeding of the SPE Annual Technical Conference and Exhibition*, Anaheim, California, USA, 11–14 November 2007. Paper SPE-109610.

Bratvold, R., Begg, S., & Campbell, J. (2002). Would you know a good decision if you saw one? *Society of Petroleum Engineers*. In *Proceeding of the SPE Annual Technical Conference and Exhibition*, San Antonio, Texas, USA, 29 September–2 October 2002. Paper SPE-77509.

Bratvold, R., Bickel, J., & Lohne, H. (2007). Value of information in the oil and gas industry: Past, present, and future. *Society of Petroleum Engineers*. In *Proceeding of the SPE Annual Technical Conference and Exhibition*, Anaheim, California, USA, 11–14 November 2007. Paper SPE-110378.

Cunha, J. (2007). Recent developments on application of decision analysis for the oil industry. *Society of Petroleum Engineers*. In *Proceeding of the SPE 2007 International Oil Conference and Exhibition in Mexico*, Veracruz, Mexico, 27–30 June 2007. Paper SPE-108703.

Cunha, J., Demirdal, B., & Gui, P. (2005). Use of quantitative risk analysis for uncertainty quantification on drilling operations-review and lessons learned. *Society of Petroleum Engineers.* In *Proceeding of the SPE Latin American and Caribbean Petroleum Engineering Conference*, Rio de Janeiro, Brazil, 20–23 June. Paper SPE-94980-MS.

Demirmen, F. (2001). Subsurface appraisal: The road from reservoir uncertainty to better economics. *Society of Petroleum Engineers.* In *Proceeding of the SPE Hydrocarbon Economics and Evaluation Symposium*, Dallas, Texas, USA, 2–3 April 2001. Paper SPE-68603.

Hansson, S. (1994). *Decision theory. A brief introduction.* Department of Philosophy and the History of Technology, Royal Institute of Technology. E-book 94 pages. https://people.kth.se/~soh/dec isiontheory.pdf.

Howard, R. (1966). Decision analysis: Applied decision theory. In *Proceedings of the Fourth International Conference on Operational Research* (pp. 97–113). Wiley-Interscience.

Howard, R. (1968, September). The foundations of decision analysis. *IEEE Transactions on Systems Science and Cybernetics.* 4(3), *1968,* 211–219.

Hurst, A., Brown, G., & Swanson, R. (2000). Swanson's 30-40-30 rules. *AAPG Bulletin, 84*(12), 1883–1891.

Luce, R., & Raiffa, H. (1957). *Games and decisions.* Wiley.

Mintzberg, H., Raisinghani, D., & Theoret, A. (1976, June). The structure of unstructured decision processes. *Administrative Science Quarterly, 21*(2), 246–275. Published by Sage Publications, Inc.

Peake, W., Abadah, M., & Skander, L. (2005). Uncertainty assessment using experimental design: Minagish Oolite reservoir. *Society of Petroleum Engineers.* In *Proceeding of the 2005 SPE Reservoir Simulation Symposium*, Houston, Texas, USA, 31 January–2 February. Paper SPE-91820-MS.

Rodrigues, L., Cunha, L., Chalaturnyk, R., & Cunha, J. (2006). Risk analysis for water injection in a petroleum reservoir considering geomechanical aspects. *Society of Petroleum Engineers.* In *Proceeding of the SPE Eastern Regional Meeting*, Canton, Ohio, USA, 11–13 October. Paper SPE-104551-MS.

Taghavifard, M., Damghani, K., & Moghaddam, R. (2009). *Decision making under uncertainty and risky situations.* Society of Actuaries (SOA). https://www.soa.org/globalassets/assets/files/resour ces/essays-monographs/2009-erm-symposium/mono-2009-m-as09-1-damghani.pdf

Vanegas, J., Cunha, J., & Cunha, L. (2005). Uncertainty assessment using experimental design and risk analysis techniques applied to offshore heavy oil recovery. *Society of Petroleum Engineers.* In *Proceeding of the SPE International Thermal Operations and Heavy Oil Symposium*, Calgary, Alberta, Canada, 1–3 November. Paper SPE-97917-MS.

Virine, L., & Rapley, L. (2003). Decision and risk analysis tools for the oil and gas industry. *Society of Petroleum Engineers.* In *Proceeding of the SPE Eastern Regional/AAPG Eastern Section Joint Meeting*, Pittsburgh, Pennsylvania, USA, 6–10 September. Paper SPE-84821.

Von Neumann, J., & Morgenstern, O. (1944). Theory of games and economic behavior. Princeton University Press.

Wu, G., Zhang, J., & Gonzalez, R. (2004). Decision under risk. In D. J. Koehler & N. Harvey (Eds.), *Blackwell handbook of judgement and decision making* (pp. 399–423). Blackwell Publishing.

# Chapter 2
# Utility Theory

**Objective**

Due to uncertainties, many outcomes for the same problem are possible. This chapter introduces the expected value concept for weighting several outcomes and producing one average result; In this chapter, the need to incorporate the decision-maker's risk attitude in the valuation process is introduced. We discuss the utility function for representing the decision-maker's risk attitude and describe several risk attitude behaviours. The concept of a rational decision is framed based on von Neumann and Morgenstern's theory. The axiomatic theory for decision making, based on Savage's theory, is explored; violations to this theory and justification for the development of Prospect theory are described.

## 2.1  Introduction to Utility Theory

For decisions concerning investment, from participating in the cheapest lotteries to getting insurance for our properties to invest in billion-dollar oil and gas projects, decision-makers must weigh between the risk of the investment and its profitability because both need to be incorporated in the decision-making process.

Several elements of the theory of decisions were developed in the eighteenth and nineteenth centuries or even earlier. This period also corresponds to the formulation of probability theory. However, the mathematical foundations of the theory of decisions were not developed until the 1940s and 1950s through two remarkable works: von Neumann and Morgenstern (1944), *Theory of Games and Economic Behavior*, and Savage (1954), *The Foundations of Statistics*. They develop the axiomatic foundation for contemporary decision theory.

The English philosopher, jurist and social reformer Jeremy Bentham (1748–1832) and the Scottish historian, economist, political theorist, and philosopher James Hill (1773–1836) state that human action aims to seek pleasure and avoid pain. According to these ideas, every object or action may be considered the source of pleasure or pain, which measures the object's utility: pleasure means giving positive utility, and pain means giving negative utility. Thus, the goal of an action is to provide maximum utility.

In the decision-making framework, people choose, from all possible alternatives, the one that leads to the highest excess of positive over negative utility. Thus, the notion of utility maximisation is the fundamental part of utility theory.

Hansson (1994), following Schoemaker (1982), describes expected utility as the dominant paradigm in decision making since the Second World War.

The concept of *expected value* (EV) was introduced by the Dutch physicist, mathematician, astronomer, and inventor Christiaan Huygens (1629–1695) in his book *De Ratiociniis in Ludo Aleae* (1657). EV is the weighted average of the values where the weighting factors are the probability assigned to each outcome. Values represent a measure used to characterise the worth to the decision-maker. Years later, Jacques Bernoulli shows (*Ars Conjectandi*, 1713) that the result of EV calculation is valid if the same wager is repeated many times (law of large numbers).

When the concept of EV is extended to utility values, it is called *expected utility value* or *expected utility* (EU). When used with money, it is called *expected monetary value* (EMV).

As observed in Chap. 1, EMV is a method extensively used to decide problems under uncertainty. However, it could lead to contradictory results, as we show in Example 2.1.

**Example 2.1: The Expected Monetary Value** Between the two games posed in Table 2.1, which one do you prefer?

Using the definition of EMV, Game 2 has a much higher EMV than Game 1; however, most people, but not all, would select Game 1 over Game 2 because the risk associated with Game 2 is much higher.

The expected value approach makes decisions based on long-run averages, e.g., it considers the average value over a large number of similar games. Thus, for example, when we play ten times Game 1 described in Example 2.1, we can lose a maximum of $2,000, while playing Game 2 ten times, we can lose up to $1,000,000. But, of course, different people have different appetites to risk losing those different amounts of money that have the same likelihood to occur.

People often make decisions looking at the amount they can lose; in Game 2, there is a 50% probability of losing $100,000 in one trial, which can be a significant figure depending on the level of wealth.

**Table 2.1** Example 2.1: expected value

|        | Win, $    | Probability | Expected monetary value |
|--------|-----------|-------------|-------------------------|
| Game 1 | 2,500     | 0.5         | 1,150                   |
|        | −200      | 0.5         |                         |
| Game 2 | 250,000   | 0.5         | 25,000                  |
|        | −100,000  | 0.5         |                         |

A decision-maker that uses EMV is indifferent between a lottery defined by a 50/50 chance of gaining $100 and losing $50 and another lottery determined by a 50/50 chance of earning $5,000 and losing $4,950; that happens because both lotteries have the same EMV = $25. However, in the first lottery, we can lose just $50, but in the second one, we can lose $4,950. If your total wealth is significant (e.g., millions of dollars), you may take the risk of a substantial loss and even $4,950 may seem insignificant, but if your total wealth is small (e.g., 000), you may not take the risk of losing $4,950. So, different financial circumstances of two individuals may lead to different decisions, and the EMV criterion does not capture these differences.

Utility theory or risk preference theory is an extension of the EMV concept that includes the risk attitude of the decision-maker.

## 2.2 Historical Development of the Utility Theory Concept

The Swiss mathematician, statistician, physicist, and physician Daniel Bernoulli (Groningen, 1700, Basel, 1782) discusses in his paper *Specimen Theoriae Novae de Mensura Sortis* (1738) or *Exposition of a New Theory on the Measurement of Risk* the following lottery problem.

**Example 2.2: Bernoulli Lottery** A poor man finds a lottery ticket that has an equal probability of gaining 20,000 ducats or nothing. Can he consider selling the ticket for 9,000 ducats?

Table 2.2 shows the decision matrix of this example,

EV (keep the lottery ticket) = 10,000 ducats

EV (sell the lottery ticket) = 9,000 ducats

Then, the optimum decision is to keep the lottery ticket. However, many people agree that most poor men will sell the ticket because of the risk associated with losing all.

However, a rich man, someone with a wealth of millions of ducats, likely will decide to keep the lottery ticket for the expectation to get the 20,000 ducats.

This example suggests that value (or monetary value) may not always be the way to make decisions, as indicated in the previous section.

The famous *Saint Petersburg paradox* exemplifies the observation that people express preferences unrelated to money; this paradox has been presented in several ways. Here we discuss one of those forms.

**Example 2.3: Saint Petersburg Paradox** Let us assume that there are two players, A and B, and that player A should pay to player B an agreed participation fee to

**Table 2.2** Bernoulli lottery ticket

|  | The ticket wins | The ticket loses |
|---|---|---|
| Keep the lottery ticket | 20,000 | 0 |
| Sell the lottery ticket | 9,000 | 9,000 |

**Table 2.3** Saint Petersburg
paradox

| Outcome | Probability of occurrence | Reward to player A ($) |
|---|---|---|
| T | 1/2 | 2 |
| HT | 1/4 | 4 |
| HHT | 1/8 | 8 |
| HHHT | 1/16 | 16 |
| HHHHT | 1/32 | 32 |
| HHHHHT | 1/64 | 64 |
| H n times then T | $1/2^n$ | $2^n$ |

play a lottery that consists of tossing a fair coin which can produce two mutually
exclusive outcomes: head (H) or tail (T); the rule for this lottery is that it will stop
when the first tail appears, and then the player A will receive the amount of $\$2^n$
where $n$ is the number of tosses observed when the first tail turns up. Player A can
participate in this lottery as many times as player A wishes, each time paying the
agreed participation fee. Table 2.3 shows the first few possible outcomes with their
corresponding probabilities and rewards.

The expected monetary value that player A estimates for this lottery is,

$$EMV_A = \left(\frac{1}{2}x\$2\right) + \left(\frac{1}{4}x\$4\right) + \left(\frac{1}{8}x\$8\right) + \left(\frac{1}{16}x\$16\right)$$
$$+ \left(\frac{1}{32}x\$32\right) + \cdots + \left(\frac{1}{2^n}x\$2^n\right) - \$1 \qquad (2.1)$$

This expression is a series of $n$ terms, whose result is,

$$EMV_A = (\$1) + (\$1) + (\$1) + (\$1) + (\$1) + (\$1) + \cdots \left(\$\frac{2^n}{2^n}\right) - \$1 \qquad (2.2)$$

When $n$ grows to infinity, this series has infinite value, meaning that player A
would be willing to pay an infinite amount of money to participate in this lottery
based on the expected monetary value criteria.

This assessment means that player A will pay whatever amount requested for
playing the lottery and will be willing to participate as many times as possible.
However, Player B will never accept this lottery, independently of the amount of
money offered by player A.

However, is this a sensible result? Observing Table 2.3, the probability of winning
more than $8 is 12.5%, more than $32 is 3.1%, and more than $128 is only 0.8%.
Thus, not many people will be ready to pay more than a few hundred dollars to
participate in this lottery.

The Swiss mathematician Nicolas Bernoulli (1695–1726) proposed this problem in 1713, and it is named the Saint Petersburg paradox; this paradox arises by using the expected monetary value as a decision criterion.

The Saint Petersburg paradox (named after the journal's name where it was published) was solved by Bernoulli (1738); he argued that individuals care not for money but the utility derived from the money received. This approach was novel at Bernoulli's time because it was against the prevalent thinking. The dominant thought derived from Blaise Pascal (France, 1623–1662, mathematician, physicist, theologian, philosopher and writer) and Pierre de Fermat (France, 1601–1665, lawyer, mathematician) is that the amount of money for each alternative should be the comparison factor for deciding between other options. They supported the idea that the value of a lottery should be equal to everybody and that it must be estimated from the EMV.

Bernoulli developed the concept that the individual preference for money is inversely proportional to the person's wealth; therefore, the utility for money follows a mathematical function that is concave downward. Bernoulli replaces the maximisation of the monetary value with the maximisation of the utility value. He rejects expected value as a criterion for making decisions under risk. He introduces the utility value, a subjective function that shows that people with different desires and wealth levels value a lottery differently.

The justification for using EU instead of EMV is that two people with different desires and wealth levels value a lottery differently. The utility that a person gives to wealth does not increase linearly with the amount of money but instead grows at a decreasing rate.

The EMV is given by,

$$E(x) = \sum_{i=1}^{n} p_i x_i \tag{2.3}$$

The EUV is calculated using,

$$E(U(x)) = \sum_{i=1}^{n} p_i U(x_i) \tag{2.4}$$

In the case of the Saint Petersburg paradox, Bernoulli used a logarithmic utility function, which is a concave function, and in this case, the expected utility value is,

$$E(U(x)) = \sum_{i=1}^{n} p_i ln(x_i) \tag{2.5}$$

This series becomes a finite sum to estimate the value that player A would be willing to pay to participate in the lottery. Bernoulli argued that people maximise their expected utility from a gamble rather than the expected monetary value.

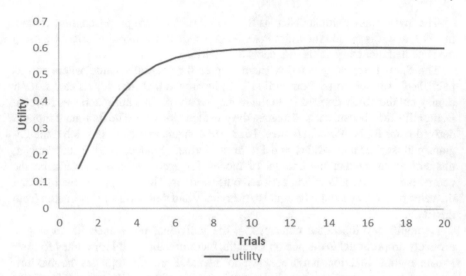

**Fig. 2.1** Saint Petersburg paradox utility curve

Expected utility theory is a normative theory that includes the risk attitude of the decision-maker and the relative values for incremental increases in money.

Figure 2.1 shows the behaviour of a concave utility function (logarithm) as the numbers of trials increase, corresponding to the Saint Petersburg paradox.

Because of the concavity of the utility function, the utility is not linear with the value (trials are linear with values). Thus, for this function, utility reaches an asymptotic value of 0.602; this utility value corresponds to the value of $4, which turns out to be a good approximation of what a typical person will be willing to pay for this lottery.

Figure 2.2 shows, in the same plot, the difference between the EMV and the EUV for the Saint Petersburg paradox.

We can see that while the expected value is a linear function that increases linearly with the number of trials, the expected utility value shows an asymptotic behaviour reaching the maximum value of 0.602.

The Saint Petersburg paradox shows that EMV is not always (indeed most of the time is not) an adequate measure to value a decision problem and indicates that decisions should be made using, instead, EUV. Most human beings assess the value of an action not based on wealth but on the consumer goods or satisfaction that can be achieved with wealth. Of course, there should be a relationship between wealth and satisfaction, which is not necessarily a linear function; this relationship between every wealth level (objective) and the level of comfort (subjective) or utility received by the individual with this wealth is characterised by utilising the utility function.

To better understand the impact of using utility instead of value for making decisions, we can estimate the difference between them: ten times increase in value from $10 to $100 corresponds, using a logarithmic utility function, to two times increase

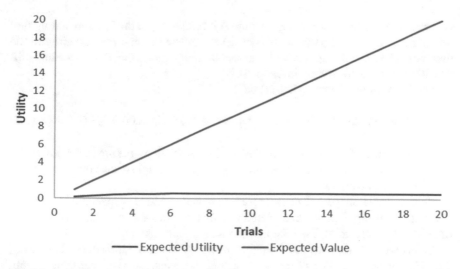

**Fig. 2.2** Saint Petersburg paradox: value and utility

from 1 to 2; in this way, the utility derived from the first $10 is equivalent to the utility derived from the next $90.

The function that transforms from wealth (money) to utility exhibits two basic properties of "rational behaviour":

(i)   a higher value of wealth should correspond to a higher value of utility; the utility function is an increasing function of wealth, and

(ii)  the marginal utility of wealth is decreasing with wealth, or the function is concave, which is mathematically represented by a twice-differentiable function with a negative second derivative. This latter property means that an increase in value is more valuable when the level of wealth is lower than when it is higher.

The following example can explain these properties of the utility function.

**Example 2.4: Diversification** Case A: An individual has their total wealth split into two parts: $2,000 in a savings account (no risk) and $2,000 in an investment in the market. The individual knows that the probabilities are 50% that the investment goes well and 50% that the investment goes wrong; in the case that the venture goes well, the individual will get $2,000 in benefits from the original investment (total $4,000), and in the case of the investment goes wrong the individual will lose the investment (total $0).

The EMV of this lottery (Case A) is

$$EMV = 0.5 * \$2{,}000 + 0.5 * \$6{,}000 = \$4{,}000 \tag{2.6}$$

Case B: The individual considers that there could be safer ways to invest: the individual believes diversification can provide a better project, and for that, the individual decides to keep $2,000 in a saving account (no risk) but split the investment (another $2,000) into two equal parts ($1,000 each) and make the first investment with the first part. Then, the investment with the second half expects that that procedure will provide a better value for the investment.

The EMV of this lottery (Case B) is:

$$EMV = 0.25 * \$2,000 + 0.5 * \$4,000 + 0.25 * \$6,000 = \$4,000 \qquad (2.7)$$

The outcome of these two expected value calculations means that both cases, against the individual's perception, have the same value; this means that diversification does not add any value.

This example suggested to Bernoulli that mathematical expectation is not an adequate measure of an uncertain lottery or investment value and supports the idea of using EUV instead of EMV for making uncertain decisions.

In this way, what matters to the decision-maker is not the monetary value. Still, the utility value or degree of satisfaction that the decision-maker can obtain from the outcomes resulting from the decision depends on the decision maker's preferences. Thus, the relationship between wealth and utility is defined through the utility function.

In the cases A and B just discussed above and using the logarithmic utility function, the expected utility values for case A is,

$$EUV = 0.5 * 3.30103 + 0.5 * 3.77815 = 3.53959 \qquad (2.8)$$

and for case B, the expected utility value is,

$$EUV = 0.25 * 3.30103 + 0.5 * 3.60206 + 0.25 * 3.77815 = 3.57083 \qquad (2.9)$$

These results show that case B is better than case A according to a decision-maker whose preferences are captured by a logarithmic utility function. Mathematically, that happens because the logarithmic function is concave. The diversification case generates a distribution around the $4,000 value, which is linear in the expected wealth but is concave in the expected utility, and the following inequality holds.

$$U(6,000) - U(4,000) = 0.17609 < 0.30103 = U(4,000) - U(2,000) \qquad (2.10)$$

This inequality occurs because of the non-linearity of the utility function.

## 2.3   Risk Attitude

In terms of *risk attitude*, a decision-maker can be *risk-averse* (the most common), *risk-seeking or risk-neutral*.

A risk-averse decision-maker would prefer to trade a gamble for a certain amount less than the EV of the venture. A classic example of a risk-averse attitude is acquiring an insurance policy; most people are risk-averse.

A risk-seeking individual is willing to pay for a gamble an amount that is higher than its expected value. A classic example of a risk-seeking decision-maker is an individual playing a state lottery with a negative expected value.

A risk-neutral individual pays for a gamble the same amount as the EV of that lottery; when a decision-maker is risk-neutral, the EMV and the EUV are the same; a risk-neutral individual is said to be indifferent to risk. Thus, mathematically, a risk-neutral decision-maker follows Eq. 2.11,

$$E(U(x)) = E(x) \tag{2.11}$$

A risk-averse utility function is a twice-differentiable function $U(x)$ satisfying two properties: (i) the first derivative is larger than zero, $U'(w) > 0$, and (ii) the second derivative is negative, $U''(x) < 0$.

The first property secures that more wealth is always preferred than less wealth (it is a monotonically increasing function). The second property (the risk-averse condition) states that the utility function is concave or the marginal utility of wealth decreases as wealth increases.

The marginal utility of money relates to the observation that $10 means more for someone whose abundant wealth is $100 than for someone whose abundant wealth is $1,000,000.

A risk-averse decision-maker is characterised by preferring to keep, for any value of health, a sure $\underline{Y}$ value than the lottery with 0.5 chance of winning $\underline{h}$ and 0.5 chance of losing $\underline{h}$:

For this preference to occur, the utility function must have the shape described in Fig. 2.3.

However, a typical decision-maker has more complex utility functions; indeed, the decision-maker may be risk-seeking for a specific range of values, risk-neutral for another range and risk-averse for a different range. For example, such utility functions explain that one individual purchases insurance and a lottery ticket simultaneously. Figure 2.4 represents these three ranges in the same decision-maker.

Figure 2.4 shows that an individual exhibits, simultaneously, several risk attitudes:

(i)   risk-averse attitudes (concave utility function for some levels of wealth, typically large values) such as when the individual buys insurance,

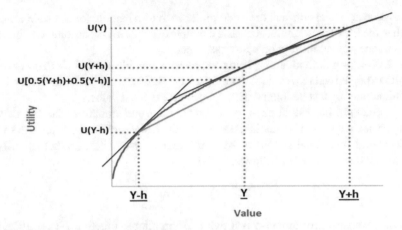

**Fig. 2.3** Risk-averse utility function

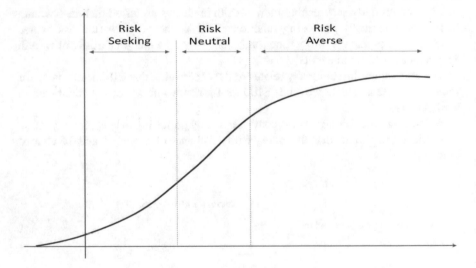

**Fig. 2.4** Utility function for a standard decision-maker

(ii)    risk-seeking attitudes (convex utility function for other wealth levels, typically for low values) when the individual buys a lottery ticket, and

(iii)   risk-neutral attitudes (linear utility function, typically for medium values).

In the field of decision theory, the marginal utility of profit is defined by,

$$MU_{profit} = \frac{\Delta U(x)}{\Delta x} \tag{2.12}$$

where $U$ is the utility function, and $x$ is a value. For a decision-maker, when the marginal utility of profit is diminishing, we say that the decision's maker's attitude is risk-averse. When it is increasing, we say that the decision maker's attitude is risk-seeking. When it is constant, the decision-maker's attitude is risk-neutral. Of course, the ranges of values for the different risk attitudes are person dependent.

Exponential utility function

We can often assume that a given function can represent the risk attitude of the decision-maker. One of the most used functions is the Exponential function, described as per Eq. (2.13):

$$U(x) = 1 - \exp(-x/R) \tag{2.13}$$

where $R$ is called the risk tolerance factor and $U(x)$ is a concave function representing a risk-averse attitude that approaches 1 for very large values of $x$. The utility of zero is zero.

For large values of $R$ the function is flatter, and for smaller values, it is more concave, has higher curvature and represents a more risk-averse attitude. Also, for less risk aversion, R should be a larger value than for more risk-averse.

In practical applications, the value of R can be determined as the highest value of $Y$ for which you will be willing to take the bet:

Win $Y with a probability of 0.5.
Lose $Y/2 with a probability of 0.5.

The value of R can be estimated by answering the question: how much will you be willing to risk ($Y/2$) to triple your money? The highest value of $Y$ for which you would prefer to gamble rather than take your money is your risk tolerance.

## 2.4  Positive Affine Transformation

A real-valued function $U$ on a linear space $L$ is affine if it satisfies:

$$U\big(\alpha l + (1 - \alpha)l'\big) = \alpha U(l) + (1 - \alpha)U\big(l'\big) \tag{2.14}$$

for all $l, l' \in L$ and $\alpha \in [0, 1]$.

$U$ is unique up to positive affine transformation if and only if any other real-valued and affine function $V$ on $L$ satisfies,

$$V(x) = \beta U(x) + \gamma \tag{2.15}$$

where $\beta > 0$.

$$V(x) = \beta U(x) + \gamma \Rightarrow U(w) \geq U(y) \Leftrightarrow V(w) \geq V(y) \tag{2.16}$$

Besides, these transformations preserve the same preferences under uncertainty:

$$E[U(x)] \geq E[U(y)] \Leftrightarrow E[V(x)] \geq E[V(y)] \tag{2.17}$$

From the above definition of affine transformation, we conclude that the utility function generated by an affine transformation of a risk-averse utility function is also risk-averse.

**Example 2.5: Affine Transformation of Utility Functions** The square root function is a risk-averse utility function,

$$\sqrt{w} = u \tag{2.18}$$

where $w$ is the value, and $u$ is the utility value.

The utility functions can be scaled by multiplying by a positive constant and/or translated by adding a constant number; such a transformation is called a *positive affine transformation*.

Examples of affine transformation over the above utility function are,

$$7\sqrt{w} \tag{2.19}$$

$$5\sqrt{w} + 40 \tag{2.20}$$

In these cases, the shape is the same but scaled and translated with respect to the original function. The preferences are the same for utility functions related by an affine transformation (Fig. 2.5).

## 2.5   Certainty Equivalent

The certainty equivalent (CE) is the sure amount of money the decision-maker is willing to pay to participate in a lottery, irrespective of the expected monetary value. CE is the value that the decision-maker assigns to the lottery, which is always lower than the expected value of the lottery for a risk-averse decision-maker. In other words, the decision-maker will be willing to sell or forgo their participation in the lottery if the decision-maker receives a proposal for an amount higher than the CE. Still, the decision-maker will be reluctant to sell the lottery if the offer is less than the CE.

The risk premium is the difference between the expected value of the lottery and the CE that the decision-maker assigns to the lottery; in other words, the risk premium is the amount the decision-maker will pay to avoid the risk.

**Example 2.6: Decision with a Square Root Utility Function** Let us assume the decision-maker has a total wealth of $200. Then, we conjecture that the decision-maker's risk attitude can be represented with the square root function:

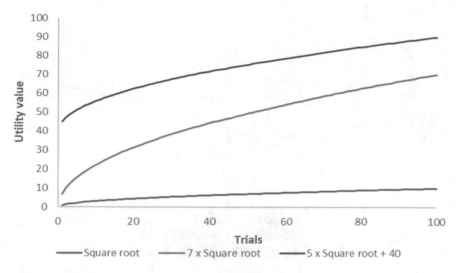

**Fig. 2.5**  Affine transformation of a square root utility function

$$U(w) = \sqrt{w} \tag{2.21}$$

Suppose that the decision-maker receives the option to participate in the following lottery:

$$\left[ \begin{array}{l} \text{Win \$200 with a probability of 0.50,} \\ \text{Lose \$150 with a probability of 0.50.} \end{array} \right.$$

The EMV of this lottery is $225.0.

The utility value in winning is 20.0, and the utility value in the case of losing is 7.1; consequently, the EUV is 13.5.

In this lottery, the EMV is positive, which suggests that for a neutral decision-maker, the decision should be to accept the lottery, which is also the assessment using EUV.

Figure 2.6 shows, using the figures of this example, how all the above quantities are related.

CE can be used as a means for ranking projects. Instead of using EMV, CE also includes the risk associated with the decision-maker and has the advantage over the utility value to use the value units. Two alternatives with the same CE have the same expected utility, and consequently, the decision-maker is indifferent to both.

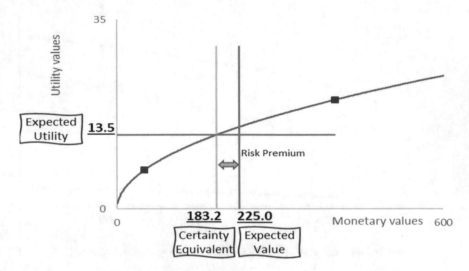

**Fig. 2.6** Certainty equivalent, risk premium, expected value and expected utility of Example 2.6

## 2.6 Formulation of Utility Theory

There are two different meanings that people use for the concept of probability:

(i)   The aleatory or random meaning of probability is also called *"objective"* probability: This meaning understands the probability of events as the relative frequency of those events in repeated trials. In this approach to probability, we include the classical meaning (Bernoulli, Laplace) that assumes the probability of an event as the ratio of the number of cases for a given outcome to the total number of cases and the frequentist meaning, which is the frequency of occurrence of a given outcome for repeated number (could that be infinite) of trials.

(ii)  The epistemological meaning of probability also called *"subjective"* probability: This meaning understands the probability of events as the degree of belief a person has on the likelihood for those events to occur. In this approach to probability, we include the logical meaning (Keynes) that assumes that we can estimate the probability of an event from the truth value of the premises of the statements related to them and the subjective meaning (Savage, De Finetti) that assumes probabilities represent a personal degree of beliefs.

Combining the *objective* and *subjective* probability approaches with the utility function gives origin to two assessment processes: decision-making under risk and decision-*making under uncertainty*, respectively.

Expected utility theory is a procedure (not the only one) used in decision theory for making choices under objective or subjective uncertainties. In the case of objective uncertainties, we call that procedure *expected utility under risk*, and for subjective probabilities, we call it *expected utility under uncertainty*.

For decision theory applications on financial problems, the utility is defined by a subjective function selected by the analyst to represent the risk attitude of the decision-maker. However, in risk analysis, the standard practice is to use objective utility; objective utility provides a base for comparing risks.

Utility theory provides the tools for constructing a subjective utility function for the decision-maker. The utility function represents the decision maker's attitude towards risk, and the best alternative is the one that maximises the expected utility.

## 2.7  Utility Theory: Decision Making Under Risk

Von Neumann and Morgenstern's theory

Bernoulli's ideas about utility values and expected utility successfully solved the Saint Petersburg paradox by incorporating new concepts. However, it did not provide their theoretical foundation; those conceptual bases were formulated by von Neumann and Morgenstern (1944) in the treatise *Theory of Games and Economic Behavior*. Von Neumann and Morgenstern provided the axiomatic system for expected utility theory.

Von Neumann and Morgenstern formulated the theory of decision making under risk or expected utility theory (EUT). EUT proposes that alternative investments should be ranked using expected utility theory, where the probabilities of the alternatives are given. Expected utility theory under risk is derived from a series of axioms that, if satisfied, provide the necessary and sufficient conditions for representing the decision-making preference under risky alternatives by a utility function.

Let us define the preference relationship $\succeq$ in the following manner.

Let ¢ be the choice set. For all $a$ and $[b \in ¢]$, $a \succeq b$ means that, according to the decision-maker's set of preferences, the decision-maker prefers $a$ rather than $b$ or at least $a$ is as preferred as $b$.

Expected utility theory under risk is based on the concept that a utility function exists with given properties. Von Neumann and Morgenstern show that the existence of the utility function can be proved if three axioms hold.

**Axiom 1: Weak-Order Axiom** The preference relationship $\succeq$ is complete and transitive.

(i)   The completeness or comparability axiom can be formulated as follows: For all $[b \in ¢]$, either $a \succeq b$ or $b \succeq a$. This axiom means that when the decision-maker chooses between two or more alternatives, those should be comparable to allow the decision-maker to decide whether one is preferred over the other or equally desirable. This axiom allows for ranking the options according to the preferences of the decision-maker. This axiom assumes that the decision-maker has well-defined preferences and the statement "I do not know which alternative I prefer" is not admissible.

(ii)    The transitivity condition can be formulated as follows: For all $[a, b, c \in ¢]$, $a \succcurlyeq b$, and $b \succcurlyeq c$ imply $a \succcurlyeq c$. This condition calls for consistency in the preferences.

**Axiom 2: Archimedean or Continuity Axioms**  These axioms impose continuity on the preference relation and should hold in both forms, A and B.

A.    *Archimedean*: For all $[a, b, c \in ¢]$, if $a \succ b$ and $b \succ c$, then there exist $\alpha, \beta \in (0, 1)$ such that,

$$\alpha a + (1 - \alpha)c \succ b \text{ and } b \succ \beta a + (1 - \beta)c \tag{2.22}$$

B.    *Continuity*: For all $[a, b, c \in ¢]$, if $a \succ b \succ c$, then there exists $\alpha \in (0, 1)$ such that,

$$\alpha a + (1 - \alpha)c \succ b \succ (1 - \alpha)a + \alpha c \tag{2.23}$$

Another way to state the *continuity* axiom is that there exists a probability $\alpha \in (0, 1)$ such that,

$$\alpha a + (1 - \alpha)c \sim b \tag{2.24}$$

This axiom means that the decision-maker is indifferent between the lottery and the sure value.

The *Archimedean* axiom says that the separation between extreme preferences can be maintained when a small deviation in the probabilities is applied: when changing the parameters $\alpha$ and $\beta$ within the interval $(0, 1)$, the choice of extreme alternatives will approximate the preference of the intermediate option. The *continuity* axiom means that there exists an inflexion point between two strict preferences. Only one of these axioms, *Archimedean* or *continuity*, should hold to satisfy von Neumann and Morgenstern's theory.

The *Archimedean* axiom is a fascinating one. Howard Raiffa proposes the following thought experiment as a counterexample: Suppose you are about to buy a newspaper that costs $1; before buying it, you notice that it is offered for free in the street on the other side. The question is: would you cross the street to get it for free even if the possibility to have an accident while crossing the street is low? If the answer is yes, meaning you would cross the street, you accept the risk of the accident for $1; otherwise, the axiom is not fulfilled. What is the value of the probability of an accident while crossing the street that makes the axiom invalid.

**Axiom 3: Independence or Substitution Axiom**  This is the most important axiom of the theory. For all $[a, b, c \in ¢]$ and $\alpha \in (0, 1]$, $a \succcurlyeq b$ if and only if

$$\alpha a + (1 - \alpha)c \succcurlyeq \alpha b + (1 - \alpha)c \tag{2.25}$$

The independence axiom imposes a separability on the preference relation; thus, the base for comparing alternatives is their distinctions. The lottery $\propto a + (1- \propto)c$ differs from the lottery $\propto b + (1- \propto)c$ in the terms $a$ and $b$ so, if $a \succcurlyeq b$, the Eq. 2.23 holds.

Based on this axiom, the preference between lotteries is the same as the parts that make them different or between compound lotteries that otherwise will be the same. This axiom is also called the *substitution axiom*, and it is the most important axiom of this theory.

In the previous axioms, each alternative is a lottery (or prospect) consisting of one or more outcomes with their respective associated probability. In general, we denote a lottery with $n$ possible results as $(x_1, p_1; \ldots; x_n, p_n)$, where the lottery produces the outcome $x_i$ with the probability $p_i$.

As discussed by Kahneman and Tversky (1979), the three principles of expected utility theory are the following:

(1) *Expectation*: The overall utility of a lottery is equal to the expected utility of its outcomes:

$$U(x_1, p_1; \ldots; x_n, p_n) = p_1 U(x_1) + \ldots + p_n U(x_n) \qquad (2.26)$$

(2) *Asset integration*: The utility of a lottery is acceptable if the utility of the integration of the lottery with any asset is larger than the utility of the asset alone; this principle means that for any asset $w$,

$$U(w + x_1, p_1; \ldots; w + x_n, p_n) > U(w) \qquad (2.27)$$

(3) *Risk aversion*: An individual is risk-averse if the individual prefers the certain lottery $(x)$ to any risky lottery with expected value $x$; this risk-averse preference is captured by a utility function with a concave-shaped function.

$$U'' < 0 \qquad (2.28)$$

## Theorem of von Neumann and Morgenstern

Von Neumann and Morgenstern proposed the necessary and sufficient conditions for representing a preference relationship on risky alternatives using an expected utility function:

Let $\succcurlyeq$ be a binary relation on $\mathcal{C}$. Then $\succcurlyeq$ satisfies the weak-order, *Archimedean* and *independence* axioms if and only if there exists a real-valued, affine function $U$ on $\mathcal{C}$ that represents $\succcurlyeq$. Moreover, $U$ is unique up to a positive affine transformation.

A monotonically increasing transformation of $U$ equally represents the preference relation.

The significance of this theorem is that it secures the existence of the utility function if the binary relation between elements of the set satisfies the three axioms of von Neumann and Morgenstern.

## 2.8   Utility Theory: Decision Making Under Uncertainty

<u>Savage's theory</u>

Savage (1954) published the book *The Foundations of Statistics*, whose major contribution was developing an axiomatic system that extends the expected utility from risk to uncertainty, creating the subjective expected utility (SEU). He proposes a decision model consisting of three sets: the set of States of Nature, $S$; the set of consequences, $C$; and the set of choices, $F$. The set of choices is made of elements called acts which are mappings from $S$ to $C$. The set of consequences contains all that could happen, and the set of states includes all possible states of the system. The combination of an act $h$ selected by the decision-maker and the state $s$ determines the consequence $c$.

Savage established a set of necessary and sufficient postulates to represent the preference relation by the expectation of a utility function on the set of consequences weighted by a probability measure on the set of all events; this utility function is unique up to a positive affine transformation.

The acts are functions from the set of *States of Nature* to the set of consequences; however, when we consider events (subset $S$), acts are conditioned to the event under which they are defined.

The act $h$, given the event $A$ and the *State of Nature* $s \in S$ is,

$$h(s) = [f, A; g, A'] = \begin{cases} f(s), & if s \in A, \\ g(s), & if s \in A' \end{cases} \tag{2.29}$$

Before discussing Savage's axioms, the following definitions are required.

**Definition 1** A constant act is an act whose consequences are independent of the state of the world.

$f \in F$ is a constant act if and only if for any $s \in S$, $f(s) = c$, where $c \in C$. $F^{const}$ is the set of constant acts.

**Definition 2** Acts are conditioned to events.

$$f \succcurlyeq_A g : if \left[ f, A; h, A' \right] \succcurlyeq \left[ g, A, h, A' \right] \text{ for a certain } h \in F \tag{2.30}$$

Equation 2.28 means that act $f$ is preferred to act $g$, when the event $A$ is given.

**Definition 3** Null event.

An event $A$ is null if $f \succcurlyeq_A g$ for any

$$f, g \in F \tag{2.31}$$

**Axiom 1: Weak-Order Axiom** The preference relation $\succcurlyeq$ is a transitive and complete binary relation on $F$.

The crucial part of this axiom is completeness, which requires the decision-maker to be able to rank any pair of acts. In that sense, the decision-maker must decide between two or more acts without ambiguity.

**Axiom 2: Sure-Thing Principle**  For any f, g, h, $h' \in F$ and non-null event $A \subseteq S$, the following holds:

$$[f, A; h, A'] \succcurlyeq [g, A; h, A'] \text{ if and only if } [f, A; h', A'] \succcurlyeq [g, A; h', A'] \quad (2.32)$$

This axiom states that the preference between two acts with a common extension outside some event $A$ does not depend on that common extension. Consequences are determined by $f$ and $g$ on the event $A$, and a common act $h$ outside $A$; the sure-thing principle says that the preference remains unchanged if we replace act $h$ by some other $h'$. This axiom is equivalent to the *Independence axiom* in von Neumann and Morgenstern's theory.

**Axiom 3: Ordinal Event Independence or Monotonicity Axiom**  For any non-null event $A \subseteq S$, and for any $f, g \in F^{const}$,

$$[f, A; h, A'] \succcurlyeq [g, A; h, A'] \text{ if and only if } f \succcurlyeq g \quad (2.33)$$

This axiom says that if an act $f$ is preferred to another act $g$, this preference is so regardless of the state in which these consequences are evaluated. In this sense, this axiom separates the *States of Nature* from the consequences.

**Axiom 4: Comparative Probability**  For any $A, B \subseteq S$ and $f, g, f', g' \in F^{const}$ such that $f \succcurlyeq g$ and $f' \succcurlyeq g'$,

$$[f, A; g, A'] \succcurlyeq [f, B; g, B'] \text{ if and only if } [f', A; g', A'] \succcurlyeq [f', B; g', B'] \quad (2.34)$$

**Axiom 5: Nondegeneracy**  There exist $f, g \in F^{const}$ such that $f \succcurlyeq g$.

**Axiom 6: Small-Event Continuity**  For any $f, g \in F$ such that $f \succcurlyeq g$ and for any $h \in F^{const}$, there exists a finite partition $P$ of the set $S$ such that for any $H \in P$:

$$\begin{array}{ll} \text{(i)} & [h, H; f, H'] \succ g, \\ \text{(ii)} & f \succ [h, H; g, H'] \end{array} \quad (2.35)$$

**Axiom 7: Dominance**  For any $f, g, h \in F$, if $f(s) \succ g(s)$ for any state $s$ of event $A$, then for any $h$,

$$[f, A; h, A'] \succcurlyeq [g, A; h, A'] \quad (2.36)$$

Savage's axioms share several similarities with the von Neumann and Morgenstern axioms. Both sets of axioms secure completeness and transitivity. They both have the sure-thing principle, which removes any common elements from the preference relationship (comparison); Axioms 3 and 4 are special for Savage's formulation to achieve subjective probability for a subjective utility.

The main advantage of Savage's theory is that it derives the subjective probabilities from axioms connected with preferences instead of assuming them a priori.

Many debates in the literature after the work of von Neumann and Morgenstern and Savage raise concerns on the validity of those theories to describe how people, in reality, make decisions. In particular, the typical situation that the same person who buys insurance also buys a lottery ticket contradicts the EU; this behaviour suggests that even if normative theories are correct as recipes to make good decisions, people make decisions following different criteria. This type of behaviour can be explained using a utility function similar to the one depicted in Fig. 2.4. The utility function has several regions of convexity, and concavity accounting for the changing risk attitude depending on the ranges of values (Friedman & Savage, 1948).

## 2.9  Bayesianism

Jacques Bernoulli (1654–1705), an uncle of Nicolas and Daniel Bernoulli, introduces the term subjective probability in his book *Ars Conjectandi* (1713); subjective probability represents the degree of belief that an event will occur; this degree of belief, of course, maybe be different with different persons.

The British mathematician, philosopher and economist Frank Ramsey (1903–1930) was the first to use subjective probabilities in his article "*Truth and probability*" (1926), where he discusses subjective probability and utility within the method of expected utility. Ramsey argues that each individual has their knowledge which results in a different probability distribution. Expected utility using subjective probabilities and subjective utilities is called Bayesian decision theory or Bayesianism.

The *Bayesianism theory* has four principles.

(1)   The Bayesian subject has a coherent set of probabilistic beliefs: compliance with laws of probability which are the same as objective probability.
(2)   The Bayesian subject has a complete set of probabilistic beliefs: each proposition has an assigned subjective probability. Thus, a Bayesian subject has a degree of belief about everything. In that sense, a Bayesian decision-maker follows a decision-making process under certainty or risk but not uncertainty or ignorance.
(3)   The Bayesian subject changes their beliefs in accordance with the subject's conditional probabilities when new evidence is presented. Let us assume that the initial probability of an event $A$ is $P(A)$ and that a new event $B$, which impacts on $A$, occurs with probability $P(B)$. Bayesianism requires that the probability of $A$ is revised according to the conditional probabilities.

$$P(A|B) = \frac{P(A \cup B)}{P(B)} \tag{2.37}$$

(4)    The Bayesian subject chooses the option with the highest expected utility.

Subjective Bayesianism (Savage and De Finetti) assumes no other rationality criteria for assigning subjective probabilities than the first three principles. However, objective Bayesianism (Jeffreys and Jaynes) believes that a unique probability assignment is admissible for the given information available to the analyst.


## 2.10  Violation of Savage's Theory and Prospect Theory

As discussed by Savage, his theory could be viewed as an immature and superficial empirical theory of foreseeing human behaviour while making decisions. Besides, human behaviour often contradicts theories, sometimes to an extraordinary degree.

Expected utility theory (EUT) in both approaches, von Neumann and Morgenstern's and Savage's, is accepted as a normative model for rational decision making and has been applied to model people's choices. Accepting EUT means agreeing with the axioms that support that theory. However, a few years after the development of Savage's model, several paradoxes show a systematic violation of those axioms.

Usually, when people make decisions, several misleading thinking styles are observed.

A.    Bias:

    i.     Recency: The most recent information can inappropriately influence people's judgement while older data are underused.

    ii.    Availability: Information is easily available to drive decisions even though they may not be the most representative.

    iii.   Vividness: Data presented vividly tend to be easier to remember for people, influencing the decision process.

B.    Overconfidence and undue optimism: Many people tend to behave with overconfidence leading to an unrealistic assessment of their ability, qualifications and knowledge, which leads to an inappropriate evaluation of their judgements and opinions.

C.    Wrong perception of reality: Many people are prone to favour evidence supporting their beliefs and dismiss evidence against their beliefs.

D.    Anchoring: When you focus on value and define the range of uncertainty close to that value, you are anchoring to that value; when a "most likely" value is set, many times people create optimistic and pessimistic values close to the "most likely" value; the anchoring effect can also occur with qualification words: "How low are the values of porosity?" versus "How large are the values of porosity?" is an example.

E.    Stop searching: Many people believe that having more data is always better
      because it reduces uncertainties; however, having more data increases our confi-
      dence but does not reduce uncertainties, and even in some cases, having more
      data decreases the accuracy of our predictions.

## Prospect theory

Kahneman and Tversky (1979) conducted a series of surveys to investigate people's
choices under uncertain decisions; they observe that in some cases, people do not
follow the principles of EUT.

   These violations of the EUT lead to a series of *"effects"* responsible for the
observed behaviours.

## Certainty effect

The certainty effect captures the fact that, usually, people overweight or overestimates
outcomes that are considered certain compared with probable outcomes, even at the
cost of producing inconsistent decisions.

   The French economist Maurice Allais (1911–2010) and the American economist
and analyst Daniel Ellsberg (1931–present) formulated corresponding paradoxes,
counterexamples of Savage's theory.

**Example 2.7: The Allais Paradox**  The Allais paradox (1953) is an example of the
certainty effect; it was re-formulated by Kahneman and Tversky for their surveys.
In problem 1, the survey participants were offered to select between two alternative
options, A and B; alternative A is a lottery with three possible outcomes while
alternative B has only one possible outcome. Table 2.4 shows the probabilities and
rewards of each alternative and the responses obtained in the surveys.

   For problem 1, most people participating in the survey prefer alternative B rather
than A (82 vs. 18).

   The same participants in problem 1 were asked for their preference in Problem 2
(summarised in Table 2.5). In problem 2, each alternative has two possible outcomes.
Table 2.5 shows, for problem 2, the probabilities and rewards and the resulting
outcomes amongst the participants in the survey.

   For problem 2, most participants prefer alternative C rather than D (83 vs. 17).

   However, as shown below, when describing the Allais paradox, the results of the
survey in problems 1 and 2 lead to contradictory inequalities for any utility function
describing the preferences of the decision-maker:

**Table 2.4**  Allais paradox: Problem 1

| Problem 1 | Probability | Value | Survey preference result (%) |
|---|---|---|---|
| Alternative A | 0.33 | 2,500 | 18 |
| | 0.66 | 2,400 | |
| | 0.01 | 0 | |
| Alternative B | 1.00 | 2,400 | 82 |

**Table 2.5**  Allais paradox: Problem 2

| Problem 2 | Probability | Value | Survey preference result (%) |
|---|---|---|---|
| Alternative C | 0.33 | 2,500 | 83 |
| | 0.67 | 0 | |
| Alternative D | 0.34 | 2,400 | 17 |
| | 0.66 | 0 | |

From problem 1,

$$0.34U(2,400) > 0.33U(2,500) \tag{2.38}$$

From problem 2,

$$0.33U(2,500) > 0.34U(2,400) \tag{2.39}$$

These examples show that selecting the preferred alternatives for these problems is contradictory between them, which results in their mutual selection to lead to irrational decisions.

The logic behind the observed selection and preference is this. In problem 1, the sure thing of alternative B (winning $2,400) makes it more attractive than the risky alternative A. In problem 2, the two probabilities (0.34 and 0.33) are very similar, so the preferred option is the one with the higher value. These results contradict EUT and can be explained as a consequence of bias.

Besides, problem 2 is obtained from problem 1 by eliminating the 0.66 chance of winning $2,400 in the first alternative and adding the 0.66 probability of winning $0 to both alternatives.

In problem 1, people prefer alternative B, which is certain, instead of alternative A which is uncertain; however, when only two uncertain options are compared, the one with the higher value is preferred.

This example shows that people are willing to pay a significant premium to avoid a small chance of receiving nothing.

Kahneman and Tversky (1979) discuss another example, described in Table 2.6, also based on Allais's work, which presents a more straightforward lottery.

**Example 2.8: Another Version of Allais Paradox** Let us consider the lottery described by Table 2.6, problem 3, consisting of two alternatives A and B,

**Table 2.6**  Allais paradox: Problem 3

| Problem 3 | Probability | Value | Survey preference result (%) |
|---|---|---|---|
| Alternative A | 0.80 | 4,000 | 20 |
| | 0.20 | 0 | |
| Alternative B | 1.00 | 3,000 | 80 |

**Table 2.7** Allais paradox:
Problem 4

| Problem 4 | Probability | Value | Survey preference result (%) |
|---|---|---|---|
| Alternative C | 0.20 | 4,000 | 65 |
| | 0.80 | 0 | |
| Alternative D | 0.25 | 3,000 | 35 |
| | 0.75 | 0 | |

Table 2.7 shows the complementary problem, problem 4, for the Allais paradox,

The survey result means that in problem 3, people prefer alternative B rather than alternative A, 80 versus 20; in problem 4, people prefer alternative C rather than D, 65 versus 35.

In terms of expected utility, alternative A has a higher expected utility than B (3,200 vs. 3,000), and alternative C has a higher expected utility than D (800 vs. 750). However, for making decisions, and based on this survey's outcomes, most people are not following the expected utility method (Savage's approach) or rational thinking.

We prove the inconsistency of the result of this survey.

If alternative B is preferred rather than A (80% vs. 20%), then,

$$U(3000) > 0.8U(4000) \tag{2.40}$$

which can be written as

$$\frac{U(3000)}{U(4000)} > \frac{4}{5} \tag{2.41}$$

On the other hand, if alternative C is preferred rather than D,

$$0.2U(4000) > 0.25U(3000) \tag{2.42}$$

which can be written as

$$\frac{U(3000)}{U(4000)} < \frac{4}{5} \tag{2.43}$$

These two inequalities are in contradiction, meaning that the expected utility theory is violated.

In this example, alternative A is transformed into alternative C and alternative B is transformed into D if the chances are multiplied by 0.25; this result contradicts the independence axiom of utility theory: problems 3 and 4 are the same except for a fixed lottery; however, people decide the opposite in both cases.

In problem 3, people prefer alternative B because it offers certainty even if the prize is lower; however, people tend to decide based on the expected value when certainty is eliminated.

**Table 2.8** Ellsberg paradox: the reward for participating in bet 1

| | Drawing a ball of each colour | | |
|---|---|---|---|
| | Red | Yellow | Green |
| Alternative 1 | $1,000 | $0 | $0 |
| Alternative 2 | $0 | $1,000 | $0 |

**Table 2.9** Ellsberg paradox: the reward for participating in bet 2

| | Drawing a ball of each colour | | |
|---|---|---|---|
| | Red | Yellow | Green |
| Alternative 3 | $1,000 | $0 | $1,000 |
| Alternative 4 | $0 | $1,000 | $1,000 |

**Example 2.9: Ellsberg Paradox** Another paradox that contradicts Savage's theory is *Ellsberg's paradox* discussed below. Let us assume you have the following two bets,

Bet 1:

Suppose there is an urn with 90 balls: 30 of the colour red and 60 yellow or green in an unknown proportion. The player draws a ball from the urn and gets the rewards according to Table 2.8.

When faced with this bet, most people select alternative 1.

Bet 2:

After participating in bet 1, the player is invited to participate in another bet, with rewards described in Table 2.9.

In this bet, most people select Alternative 4.

However, these selections contradict Savage's theory because if alternative 1 is preferred to 2, alternative 3 should be preferred to alternative 4 to satisfy the sure-thing principle. In both cases, subjects prefer betting on known probabilities rather than unknown or vague probabilities; however, these choices contradict the sure-thing principle.

Other violations to the expected theory have been recognised, such as the reflection effect when the sign of positive outcomes is reversed, making gains into losses, or the isolation effect when alternatives are divided into different components valued separately.

Reflection effect

Kahneman and Tversky discuss the *reflection effect* and the opposite risk attitude observed in many people when assessing high or low likelihood events.

In the survey conducted by Kahneman and Tversky, when people were asked what lottery they prefer between the alternatives (i) and (ii) below:

(i)    win $3,000 for sure.
(ii)   win $4,000 with 0.8 chance and win $0 with 0.2 chance.

80% selected the first option (i). The behaviour is risk-averse.

Next, when the same people were asked to choose between lotteries (iii) and (iv):

(iii)     win $3,000 with 0.002 chance and win $0 with chance 99.998.
(iv)      win $6,000 with a chance of 0.001 and win $0 with a chance of 99.999.

73% selected the first option (iii). This behaviour is risk-neutral.

When these lotteries are converted to losses, people change their risk attitudes; 92% of people prefer lottery (v) rather than (vi):

(v)       lose $4,000 with 0.8 chance and lose $0 with 0.2 chance.
(vi)      lose $3,000 for sure.

This behaviour is risk-seeking.

Also, it was observed that 70% choose option (vii):

(vii)     lose $6,000 with a chance of 0.001 and lose $0 with a chance of 99.999.
(viii)    lose $3,000 with a chance of 0.001 and lose $0 with a chance of 99.998.

This observation represents risk-neutral behaviour.

The reflection effect captures the observation that risk-averse behaviour in the positive domain translates to risk-seeking behaviour in the negative domain. Also, people tend to be risk-averse for gains and risk-seeking for losses.

All these paradoxes lead to the development of *prospect theory*, whose formulation falls outside the scope of this book.

To summarising, prospect theory defines a model for decision making that deviates from expected utility theory in three main aspects, as explained below.

(1)     People assess the value of a decision in terms of the relative value with respect to a reference value which could be the person's wealth position or the base level of the problem; the decision-maker is interested in the changes of values, not the net value, as occurs in the expected utility value.

(2)     The value function for losses is not the same as the value function for gains; people are risk-averse for gains and risk-seeking for losses. In addition, the sense of loss for a given value is stronger than the sense of gain for the same amount. Figure 2.7 shows an example of this behaviour: at the right-hand side in blue line colour the risk-averse attitude, and on the left-hand side, in a continuous blue line, the equivalent risk-seeking for negative values, in dashed green line the adjusted risk-seeking behaviour.

(3)     People weigh the outcomes with decision weights, not probabilities; typically, people overweight low-probability outcomes and underweight medium and high chances. This issue is related to the certainty effect described above.

The best reference for prospect theory is the article of Kahneman and Tversky (1979).

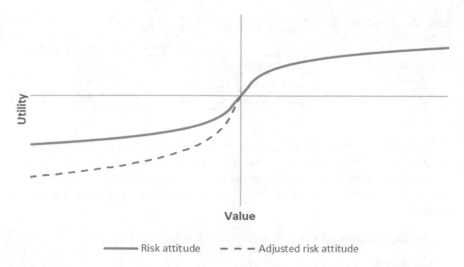

**Fig. 2.7**  Kahneman and Tversky's value function: risk-averse and risk-seeking

Decision trees

Many decision problems can be analysed using *decision trees*. Decision trees are graphical structures that organise the information relevant to the decision. A decision tree consists of:

(1)   A root node represents the beginning of the decision process.
(2)   Branches to join the root with the subsequent nodes, branches from node to node and branches from nodes to the terminal nodes.
(3)   The terminal nodes show the outcomes of the decision process.
(4)   A set of nodes between the root node and the terminal nodes: each non-terminal is either a decision node, where the decision-maker should make a decision, or a choice node, where a chance event occurs, and it is assessed according to the *States of Nature*.

The terminal node has a value or payoff, which is the value of the outcome to the decision-maker.

**Example 2.10: Decision Tree**  Let us consider an example of a decision problem that we will evaluate using a decision tree.

Consider a building development company manager dedicated to constructing houses in an area near the country's capital city. The company acquired, very recently, land with the objective to develop a plan consisting of family houses. But, first, the manager must decide what the optimum number of houses to build is. The manager considers three scenarios for construction: build 30, 60 or 90 houses; the number of houses for each scenario corresponds to a given investment level.

**Table 2.10** Payoff table for
the construction problem

| Decision alternative | State of Nature | |
|---|---|---|
| | Strong demand, $s_1$ | Weak demand, $s_2$ |
| Small complex, $A_1$ | 10 | 8 |
| Medium complex, $A_2$ | 14 | $-1$ |
| Large complex, $A_3$ | 24 | $-10$ |

Of course, the project's objective is to generate the most considerable profit possible, given the uncertainty in demand. The demand for houses is an uncertain variable. Based on previous experience and market analysis, it was decided to consider two possible scenarios for demand: strong demand and weak demand for houses.

By using decision analysis terminology, the problem can be posed as follows.

Alternatives:
$A_1$: build a small complex with 30 houses.
$A_2$: build a medium complex with 60 houses.
$A_3$: build a large complex with 90 houses.
*States of Nature*:
$s_1$: strong demand for the houses.
$s_2$: weak demand for the houses.

Whatever the decision is taken, whether to build 30, 60 or 90 houses, a consequence will occur depending on the *State of Nature*.

Based on the cost for building the houses and the price the company expects to receive for each house, we create the payoff table, which is a table showing for each alternative and *State of Nature* the associated benefit or consequence, which in this case is the profit of the investment. Table 2.10 shows the payoff table for this example.

The decision-maker controls only the decision, but the *State of Nature* is an uncontrollable variable that is impossible to know before the decision is taken unless a clairvoyant can say what will happen in the future.

The best case is that the company would build a large complex and the demand for houses will be strong; on the other hand, the worst case is that the company would build a large complex, but the demand will be weak, in which case many houses will not be sold. Therefore, the investment of building those will be a loss. In between these two extreme cases lie the remaining cases. However, at the time of making the decision to build the houses, the manager of the company does not know if the demand would be strong or weak once the houses are ready; the most that the manager can do is an estimate, based on the manager's experience (and the manager's assessors' opinion) and market surveys and studies, of what could be the probabilities that, in the future, the demand of the houses would be strong or weak.

Let us assume that, based on the manager's experience and market surveys, the manager estimates the following probabilities for the *State of Nature*. There is a 0.6

chance that the demand for houses will be strong, and there is a 0.4 chance that the demand for houses will be weak.

Applying the definition of expected value,

$$EV(A_1) = 10 * 0.6 + 8 * 0.4 = 9.2 \tag{2.44}$$

$$EV(A_2) = 14 * 0.6 + (-1) * 0.4 = 8.0 \tag{2.45}$$

$$EV(A_3) = 24 * 0.6 + (-10) * 0.4 = 10 \tag{2.46}$$

The expected value of these alternatives indicates that the best choice is alternative 3, which has a higher expected value. However, if the probabilities were opposite, 0.4 chance for strong demand and 0.6 chance for weak demand, alternative 1 would be the best choice (expected value of 8.8 for alternative 1, 5.0 for alternative 2 and 3.6 for alternative 3).

This calculation can be conveniently represented in a decision tree. Figure 2.8 shows the decision tree for this case.

Decision node 1 corresponds to the optimum decision that the decision-maker can make. Nodes 2, 3, and 4 are the probabilistic nodes whose values are calculated conditioned to the corresponding alternative; for each of these nodes, we have available the payoff or value of the alternative subject to the two levels of house demand selected by the decision-maker. The results of the nodes are shown in Fig. 2.9.

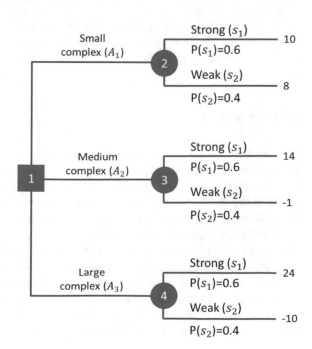

**Fig. 2.8** Decision tree for the complex development problem

**Fig. 2.9** Evaluation of
alternatives for the complex
development problem

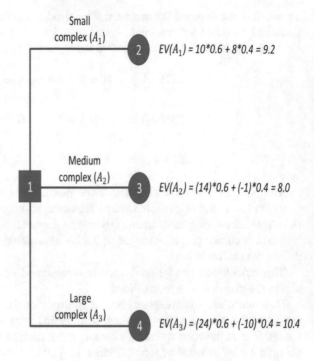

$EV(A_1) = 10*0.6 + 8*0.4 = 9.2$

$EV(A_2) = (14)*0.6 + (-1)*0.4 = 8.0$

$EV(A_3) = (24)*0.6 + (-10)*0.4 = 10.4$

Figure 2.9 is part of the decision tree shown to explain the calculations performed in this specific problem.

As shown, alternative 3 is optimum for the combination of alternatives, chances, and values estimated for this problem because the maximum expected value is 10.4, which corresponds to the alternative of building a large complex.

## 2.11  Summary

The chapter introduces the expected value concept as a critical element in dealing with uncertainty when making decisions; expected value is the standard method to value a project under uncertainty. Further, we search in history to explore the application of the concept of risk attitude to determining the value of lotteries. Several examples are cited to explain the need to use the utility instead of monetary value in valuation problems to capture the decision-maker's risk attitude; a relevant case to describe the need for using utility is the Saint Petersburg paradox. Utility functions are presented as a mathematical representation of the risk attitude, and several examples are discussed. Classification of the risk attitudes of the decision-maker is discussed and related to the characteristic of the utility function. A brief introduction to the concept of the certainty equivalent and its use in valuing and ranking decision alternatives is presented. The formulation of the utility theory by von Neumann and

Morgenstern and at a later stage by Savage are discussed and compared; these theories are the foundations of the decision-making under risk and uncertainty, respectively. A dedicated section of the chapter deals with Bayesianism and the reasons that justify it, mainly in the decision-making domain. The chapter ends with a discussion of several examples, such as Allais's paradox, which show that, even though applying Savage's theory, the rationality of a decision is secure, people frequently make decisions that contradict this theory. Understanding the reasons behind these contradictions gives birth to other decision theories, the most important being the Prospect theory discussed.

# References

Bernoulli, J. (1713). Ars Conjectandi, Opus Posthumum. Accedit Tractatus de Seriebus In-finitis, et Epistola Gallice scripta de Ludo pilae reticularis. *Basileae, Impensis Thurnisiorum*, Fratrum –We r k e3 (pp. 107–286). The Art of conjecturing. Translation by Edith Dudley Sylla, Johns Hopkins University Press, 2006.

Bernoulli, D. (1954). Exposition of a new theory on the measurement of risk. *Econometrica, 22*(1), 23–36. https://doi.org/10.2307/1909829. (Translation of Bernoulli, D. (1738). Specimen theoriae novae de mensura sortis; Papers Imp. Acad. Sci. St. Petersburg 5, 175–192.)

Friedman, M., & Savage, L. J. (1948). The utility analysis of choices involving risk. *Journal of Political Economy, 56*(4), 279–304. https://doi.org/10.1086/256692

Hansson, S. (1994). *Decision theory. A brief introduction*. Department of Philosophy and the History of Technology, Royal Institute of Technology. E-book 94 pages. https://people.kth.se/~soh/dec isiontheory.pdf

Huygens, Ch. (1657). *Libellus De Ratiociniis in Ludo Aleae OR The Value of all chances in Games of Fortune; Cards, Dice, Wagers, Lotteries*. English translation 1714. Printed by S. Keimer for T. Woodward.

Kahneman, D., & Tversky, A. (1979). Prospect theory: An analysis of decision under risk. *Econometrica, 47*(2), 263–292. https://doi.org/10.2307/1914185

Ramsey, F. (1926). Truth and probability. Published 1931 in Foundations of Mathematics and other Logical Essays, Ch. VII (pp. 156–198). Edited by R. B. Braithwaite. Kegan, Paul, Trench & Co. Ltd. Harcourt, Brace and Company.

Savage, L. (1954). *The foundations of statistics*. Wiley.

Schoemaker, P. (1982, June). The expected utility model: Its variants, purposes, evidence and limitations. *Journal of Economic Literature, 20*(2), 529–563.

Von Neumann, J., & Morgenstern, O. (1944). *Theory of games and economic behavior*. Princeton University Press.

# Chapter 3
# Probability and Inference Theory

**Objective**

Owing to the fact most decisions occur under uncertainties, and uncertainties are represented using probabilities, this chapter is dedicated to explaining the most important concepts of probability used later during the decision-making workflows. A section is dedicated to Bayes' theorem, which is used later, in Chap. 7, for developing the value of information assessment. Concepts of random sampling and statistical inference are also introduced for subsequent use in Chap. 5, experimental design.

## 3.1  Introduction to Probability

Given an event A, we say that the probability that event A will occur is P(A), where P(A) equal to zero means A will never happen, and P(A) equal to one means A will occur with certainty.

There are three definitions of the probability of occurrence:

- Relative frequency probability
- Classical or objective probability
- Subjective definition probability

Relative frequency: this definition is based on the assumption that many identical trials can be conducted. The probability of event A is defined as the number of times event A occurs divided by the total number of trials. This definition of probability is valid in the long run, where probability is established empirically through data averages. This statement does not say anything about one particular shot. The long-run concept corresponds with the frequentist interpretation of chance, associated with statistical concepts such as confidence intervals, P-values, etc.

The classic example is the toss of a fair coin, where the two possibilities are head or tail; the probability of getting a head when tossing the coin is 0.5 or 50%. The meaning of that statement is that we will get ahead in half of the times in the long run or a long sequence of identical trials.

Another typical example is when tossing a fair dice with six faces several times; if we toss the dice ten times, probably one or two faces will appear several times and others may not appear at all; however, when repeating the toss more times, all the faces will show up and after many tosses, indeed, in the long run, all faces will show the same frequency, which is $1/6 = 0.166667$ or $16.667\%$.

In exploring oil reservoir structures, this concept is not applicable because it can only apply if all the hydrocarbon accumulations in the same basin are equal, which is hardly the case.

Predicting whether a hurricane will occur at a given time within an area of frequent meteorological phenomena does not correspond with this definition of probability either. A storm is a consequence of meteorological circumstances and not a repetitive phenomenon that we can experiment with. When and where the meteorological conditions are needed for a hurricane to occur are unique occurrences and cannot feasibly be established in terms of frequency.

Classical or objective probability: this definition of probability measures the degree to which the available evidence supports a given hypothesis. This approach to probability is supported by facts, not subjective opinions. It can be applied when the data is complete, or there is a full understanding of the system. In that sense, the probability associated with an isolated outcome when a fair dice of six faces is tossed is $1/6$ without the need to toss the die hundreds of times; in the same manner, the probability that a ball thrown on a roulette wheel will land in a specific slot is equal to 1 over the number of slots.

In oil and gas exploration activities, the classical probability definition cannot be used because it assumes that we know the system, which is not the case in an exploratory oil field.

Subjective probability: this definition of probability measures the beliefs and opinions of the analyst regarding a given event occurring, and it is based on the concept of "*degree of belief*" or how likely the analyst estimates that an event will occur. This approach to probability is rooted in the work of Savage (1954). This definition of probability reflects the individual thinking, knowledge or judgment about an outcome occurring; this probability is used when no or partial data is available or when the current data is not representative of the future. This definition of probability is the basis for the development of the decision analysis theory. Individual bias, experiences, etc., impact the values assigned to the subjective probabilities. This approach to probability, although criticized, has been used successfully and extensively in areas such as decision analysis.

## 3.2 Operations with Probability

If $P(A)$ is the probability of event A occurring, $P(\sim A)$ or the probability of not A is one minus $P(A)$.

In probability theory, events such as $A$ and $B$ consist of a set of outcomes or values: $A = \{A_1, A_2, \ldots, A_n\}$ and $B = \{B_1, B_2, \ldots, B_m\}$.

Definitions:

$P(A \cup B) =$ probability of outcomes of $A$ and $B$ occurring (in set theory: intersection)

$P(A \cap B) =$ probability of outcomes of $A$ or $B$ occurring (in set theory: union)

**Example 3.1: Set Theory Operations I** If $A = \{2, 3\}$ and $B = \{3, 4, 5\}$,

$$P(A \cup B) = P(3) \tag{3.1}$$

$$P(A \cap B) = P(2, 3, 4, 5) \tag{3.2}$$

Addition theorem of probability

When we have two events $A$ and $B$, the probability of $A$ or $B$ occurring is,

$$P(A \cup B) = P(A) + P(B) - P(A \cap B) \tag{3.3}$$

This theorem means that the probability of $A$ or $B$ occurring—e.g. $P(A \cup B)$—is equal to the probability of $A$ occurring—e.g. $P(A)$—, plus the probability of $B$ occurring—e.g. $P(B)$—minus the probability of $A$ and $B$ occurring simultaneously e.g. $P(A \cap B)-$; the effect of the last term, $P(A \cap B)$, is to subtract the common elements which have been counted twice if the events are not mutually exclusive; however, when the events $A$ and $B$ are mutually exclusive, $P(A \cap B)$ is equal to zero.

**Example 3.2: Set Theory Operations II** We illustrate the addition theorem with a deck of 52 cards. Let us assume the event $A$ is drawing 1 or 2 of any suit and event $B$ is drawing 3 or 4 of any suit. For these events,

$$P(A) = \frac{8}{52} = P(B) \tag{3.4}$$

On the other hand, $P(A \cap B) = 0$ because there are no common elements between $A$ and $B$; then,

$$P(A \cup B) = \frac{16}{52} = \frac{4}{13} \tag{3.5}$$

Now, let us assume that the event $C$ is drawing 2 or 3; in this case,

$$P(A) = \frac{8}{52} = P(C) \tag{3.6}$$

but

$$P(A \cap B) = \frac{4}{52} \tag{3.7}$$

Finally,

$$P(A \cup C) = \frac{8}{52} + \frac{8}{52} - \frac{4}{52} = \frac{12}{52} = \frac{3}{13} \tag{3.8}$$

Conditioned events

Two events are dependent or conditioned when the probability of one occurring is affected by the other event that happened previously; conditional probabilities are defined by Eq. (3.9),

$$P(A|B) = \frac{P(A \cap B)}{P(B)} \tag{3.9}$$

This equation implies that the probability of $A$ occurring given that $B$ occurs is equal to the probability of $A$ and $B$ divided by the probability of $B$.

This definition makes sense because the probability of A given that B has occurred should be proportional to the joint probability of A and B occurring, with the weighting factor being the probability of B.

This definition of conditional probability can be re-arranged to generate the multiplication theorem.

Multiplication theorem of probability

When we have two events $A$ and $B$, the probability of $A$ and $B$ occurring is,

$$P(A \cap B) = P(A)P(B|A) \tag{3.10}$$

This theorem has the same information as the conditional probability, and it means that the likelihood of $A$ and $B$ occurring is equal to the probability of $A$ multiplied by the conditional probability of $B$ given that outcome $A$ has occurred $P(B|A)$.

In the case, $A$ and $B$ are independents events, the probability of events A and B occurring is given by Eq. (3.11),

$$P(A \cap B) = P(A)P(B) \tag{3.11}$$

**Example 3.3: Set Theory Operations III** Let us consider the same deck of 52 cards we discussed in the previous example; let us call event $A$ the drawing of a jack of any suit and event $B$ drawing a four of hearts. The probability of occurrence of event

$A$ and then, without replacement, event $B$, is given by Eq. (3.12),

$$P(A \cap B) = \frac{4}{52} x \frac{1}{51} = \frac{4}{2952} = \frac{1}{663} \qquad (3.12)$$

In the first draw, we have 52 cards, but in the second draw, because it is done without replacement, we have 51 cards.

Making the same draw but with replacements (once event $A$ occurs), we have,

$$P(A \cap B) = \frac{4}{52} \times \frac{1}{52} = \frac{4}{2704} = \frac{1}{676} \qquad (3.13)$$

In the last case, because the draw is with replacement, the draw is independent and the conditional probability $P(B|A) = P(B)$.

In probability, two outcomes are called mutually exclusive if the occurrence of one of them excludes the existence of the other; this concept is different from independent outcomes.

## 3.3  Population, Sample, Random Variables

A *population* is the collection of all the possible observations of a kind or all the entities of interest in a study. The characteristics of the population are called *parameters*, and they cannot be directly measured because populations are too large; they are considered fixed but unknown, and they are not random variables.

A *sample* is a collection or subset of observations of a population obtained by random sampling; the number of observations in a sample is the sample size. If there are many elements, working with a population could be impractical or even impossible; in those cases, we can work with samples, even at the cost of losing accuracy.

Measured characteristics of a sample are called *statistics,* and they are random variables because their values depend on the random sampling, and consequently, they have probability distributions. A *statistic* is a function of the observations in a sample. A particular value of a set of statistics is called an *estimate*. The statistics of the sample are used to estimate the population parameters of interest. The most common *statistics* are the ones that measure the centre of the sample, or the sample mean and the spread or the sample variance, which we discuss below in this chapter.

Good statistics should:

(A)  Be unbiased: the expected value of the statistics should be equal to the parameter being estimated.
(B)  Have a minimum variance: the statistic must be smaller than the variance of any other estimator of that parameter.

Examples of statistics are the sample means described by Eq. (3.14):

$$\overline{y} = \frac{\sum_{i=1}^{n} y_i}{n} \tag{3.14}$$

and the sample variance is given by Eq. (3.15),

$$S^2 = \frac{\sum_{i=1}^{n} (y_i - \overline{y})^2}{n - 1} \tag{3.15}$$

The sample mean is a point estimator of the population mean $\mu$, and sample variance is a point estimator of the population variance $\sigma^2$. An *estimator* of a population unknown parameter is a statistic that corresponds to that parameter.

If we know the probability distribution of a population, we can determine the probability distribution of a population statistic. The probability distribution of a statistic is called a *sampling distribution.*

A *random sample* is a sample selected from the population in such a way that every possible sample has the same probability of being selected.

A good estimator of a population parameter is one where the expected value of the sample statistics should be equal to the population parameter. In this case, repeated samples of the population should not be consistently under or over the population parameter. As the sample size increases, the estimator should get closer to the population parameter.

Given a random sample of a population, the analyst usually estimates the sample statistics to infer the parameters of the distribution corresponding to the population.

In the oil and gas business exploration activities, a sample could be the range of possible oil recoveries in the structures in a basin; a sample could also be the porosity data points measured in a core taken in a well.

A *random variable* is a parameter or variable that can have more than one value; it can be considered an experiment, process or measurement whose outcome is unknown beforehand and cannot be predicted at the time of a decision. Each of the possible values of the random variable has an associated probability of occurrence.

Random variables are also called stochastic variables to underline their probabilistic nature. A random variable is described by a probability distribution where we have the different values of the variable on the horizontal axis. In the vertical axis, we have the probability associated with the corresponding value of the variable. An outcome or event is a subset of the sample that is of interest to the analyst.

Some random variables can have only a finite or countable number of possible values described by discrete probability distributions. Discrete random variables result from counting, such as the number of prospective wells or outcomes from tossing a dice; however, discrete variables can be finite or infinite in general.

Many relevant variables in the subsurface evaluation of oil reservoirs are random variables, such as net pay thickness, field reserves, porosity values, etc.

Probability distributions

*A probability distribution* is a mathematical representation of the range and likelihood of the possible values that a random variable can have. It refers to the variables in a population; the expected value of a random variable is the mean of its probability distribution.

The mean of the distribution is a measure of the centre value. The data spread from the mean of the distribution is measured with the variance, which is the sum of the square of the deviation from the mean.

Depending on the nature of the variable, the probability distribution can be discrete or continuous, corresponding to discrete or continuous variables. The horizontal axis represents the variable (porosity, saturation, etc.) in the graphical representation of a probability distribution function. The vertical axis represents the probability associated with the corresponding value of the variable.

The probability distribution of *discrete variables* can be displayed in two ways; the first is by using the Probability Mass Function (PMF), where the probability of each outcome is represented by a bar, the height of which represents the probability of that particular outcome.

In Fig. 3.1, the binomial distribution shows the probabilities associated with the possible outcomes of tossing a fair coin 20 times, e.g., the probability of having seven heads is 0.07, as is the one for 13 heads.

For discrete probability distribution functions, the sum of the probabilities for each possible value of the random variable is precisely one.

The second way of displaying the discrete probability distribution is by using the Cumulative Distribution Function (CDF), which gives the probability that a random variable is less than or equal to a specific value: $P(X \leq x)$.

In Fig. 3.2,

$$P(X < 9) = 0.41 \tag{3.16}$$

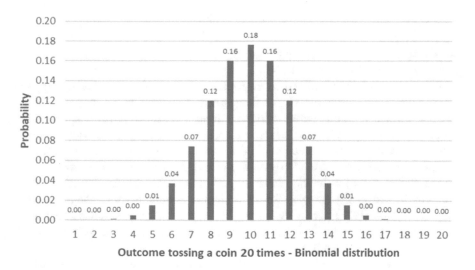

**Fig. 3.1** Binomial distribution function—Probability Mass Function (PMF)

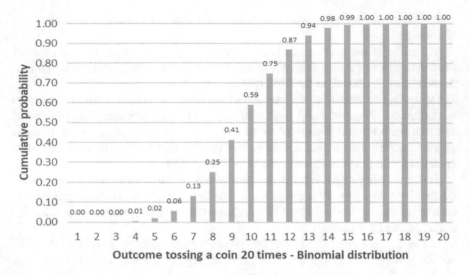

**Fig. 3.2** Binomial distribution function—Cumulative Distribution Function (CDF)

and,

$$P(X < 12) = 0.87 \tag{3.17}$$

Mathematically, for a discrete variable,

$$F(X < x_i) = \sum_{j=1}^{i} p_j \tag{3.18}$$

which means that the cumulative probability until one value, $X < x_i$, is equal to the sum of all probabilities $p_j$ until $j < i$.

Often, the variables involved in a problem are *continuous variables*, e.g., the temperature tomorrow in our city or the porosity measured in a core plug. Unlike discrete variables, for continuous variables, it does not make sense to speak of the probability at one specific value, which is zero, but a probability in intervals of values. Indeed, there are infinite numbers of values and, if each value of the variable has assigned a probability, the total probability would also be infinity. However, we can define $P(a \leq x \leq b)$ to be the probability of the variable x being between values a and b; the probability function for a continuous variable is called the *probability density function* (PDF) because we do not have a probability for each value but a probability for a range of values, or a probability per unit of the variable at any particular value of the variable.

For a continuous variable, the probability in an interval is represented by the area under the probability density function between the two extremes of the interval. Thus,

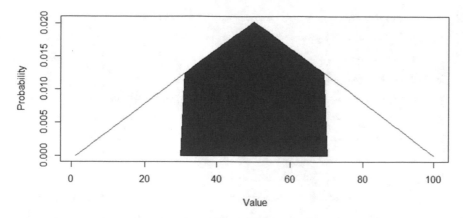

**Fig. 3.3** Triangular Probability Density Function (PDF)

the area under the probability function equals one for the complete range of values of the variables.

It is essential to realize that the probability of any particular value of a continuous variable is zero because that is the value under the curve of a single value.

Figure 3.3 shows the continuous PDF of the triangular function with minimum = 0, maximum = 100 and mid = 50 (a symmetric triangle).

In the R language, the package "Triangle" can be used to generate the triangle distribution. As shown in Fig. 3.3, the blue colour highlights the cumulative probability between 30 and 70. For this example, the area can be calculated, in R, in the following form:

$$P(30 \leq x \leq 70) = ptriangle(70, 0, 100, 50) - ptriangle(30, 0, 100, 50) = 0.64 \tag{3.19}$$

The triangle distribution is a continuous function because, for any value of x (the variable value), within $x_{min}$ (=0 in Fig. 3.3) and $x_{max}$ (=100 in Fig. 3.3), we can assign a y (the corresponding probability); the probability distribution function satisfies the property that the total area below the curve is one. Another property of the PDF is that the probabilities are always positive or zero.

When the PDF is integrated from the $x_{min}$ to any other value, we obtain cumulative probability from $x_{min}$ to the selected value; when this process is made iteratively, we obtain from the PDF the *Cumulative Distribution Function*, CDF. For example, for the Triangular PDF, the corresponding CDF is shown in Fig. 3.4.

The CDF of a continuous random variable $X$, evaluated at $x$, is the probability that $X$ takes a value less than or equal to $x$, i.e.,

$$F(x) = P(X \leq x) \tag{3.20}$$

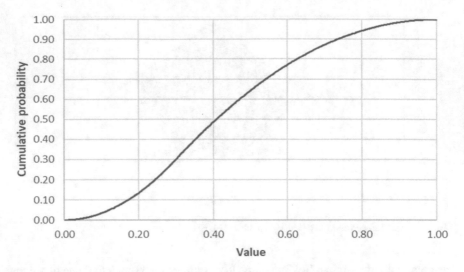

**Fig. 3.4** Triangular Cumulative Distribution Function

In a continuous distribution, the CDF is the area under the PDF curve, from the minimum value of the variable, where the probability is zero, until the maximum value of the variable with probability one.

Probability distribution descriptions

Random variables are described by the PDF associated with them. In general, those distributions are described by two parameters: (i) parameters that measure the central tendency or "average value" of the distribution, and (ii) parameters that measure the width or variability of the distribution.

Central tendency parameters

These parameters indicate how the "average" position is located along the value axis or the general location of the random variable. The most frequently used parameters to measure the central tendency are the mean, median and mode.

*Mean*: This parameter is calculated as the weighted average of the random variables where the weight factors are their probability. The mean is equivalent to the expected value of a distribution. The units of the mean are the same as the units of the random variable. The mean value is usually described with the symbol $\mu$, and it is the most critical measure of the central tendency. The mean is by far the most useful measure of the central tendency.

The sample mean $\bar{y}$ is a point estimator of the population mean $\mu$.

*Median*: This parameter is the value of the random variable where there is the same likelihood of being above or below this value; it corresponds to the 50th percentile on a cumulative distribution function. By definition, the random variable's probability being less than or equal to the mean is 0.5. This measure of central tendency is not affected by the magnitudes of the values of the random variable. If substantial values are present, the measure does not differentiate from just large values.

*Mode*: This parameter of the random variable is the most likely to occur. This value is the one that is most repeated in the sample and has the highest pick in the probability distribution. Some distributions can have more than one mode (more than one value with a higher frequency). Mode is the value we refer to when mentioning the most likely or most probable value.

In general, these measures of the central tendency give different values; however, for a symmetrical continuous probability distribution, the three parameters, mean, median and mode, provide the same value; an example of this distribution is the Normal probability distribution.

For a continuous variable, the mean value is defined as,

$$\mu = \int_{-\infty}^{\infty} x f(x) dx \qquad (3.21)$$

where $f(x)$ is the probability density function of the random variable $x$.

In the case that we have a set of data values, they can be grouped in range intervals and then by counting the number of data points in each interval; in that case, the mean is defined as,

$$\mu = \frac{\sum_i n_i x_i}{\sum_i n_i} \qquad (3.22)$$

where $n_i$ is the frequency or number of values in the interval $i$ and $x_i$ is the midpoint of the corresponding interval. This evaluation is an approximation because it uses the value at the centre of each interval to represent the several values inside that interval; however, if the size of the interval is small enough, the approximation is sufficiently accurate.

In the case of a discrete distribution, the mean value is calculated using Eq. (3.23),

$$\mu = \sum_i x_i p_i \qquad (3.23)$$

where $p_i$ is the probability of occurrence of the value of the random variable $x_i$.

Variability parameters

In addition to the measures of the central tendency for distribution, it is essential to know the dispersion or variability of the possible values of the random variable. Variability tells us whether the values are located close to the central value, near another value or dispersed around the interval range. The most critical measure of variability is the standard deviation or the squared root of the variance.

A set of values is characterized by its mean value; the difference between each independent value and its meaning is called the *deviation*; the mean value of the squared deviation is the *variance*. The squared root of the variance is the *standard deviation*. The standard deviation indicates the degree of dispersion of the data around the mean value: the higher the standard deviation, the more extensive the data spread.

Other measures of variability are the range or the 10th–90th percentile range, but these are much less useful.

The mathematical expression for calculating the *standard deviation* is given by Eq. (3.24),

$$\sigma = \sqrt{\frac{\sum_{i=1}^{N}(x_i - \mu)^2}{N}} \tag{3.24}$$

where $x_i$ are the values of the random variable, $\mu$ is the mean value, and $N$ is the number of values of the population.

When dealing with a large set of data points, and similarly to the mean value calculation, an approximation of the standard deviation can be calculated by dividing the range of values into intervals and calling $n_i$ the number of data points in the interval $i$; $x_i$ is the midpoint value of the same interval; then,

$$\sigma = \sqrt{\frac{\sum_i [n_i (x_i - \mu)^2]}{\sum_i n_i}} \tag{3.25}$$

where $n_i$ is the number of data points in the interval $i$ and $x_i$ is the midpoint value of the same interval.

Equations (3.24) and (3.25) can be re-arranged in the following easy-to-work form:

$$\sigma = \sqrt{E(x^2) - (E(x))^2} \tag{3.26}$$

which is the notation of the standard deviation, using the expected value operators.

In the case of a continuous variable, the standard deviation is defined by Eq. (3.27),

$$\sigma = \sqrt{\int_{-\infty}^{\infty} (x - \mu)^2 f(x) dx} \tag{3.27}$$

where $f(x)$ is the probability density function.

For a discrete variable, the standard deviation can be calculated using Eq. (3.28),

$$\sigma = \sqrt{\sum_i p_i (x_i - \mu)^2} \tag{3.28}$$

where $p_i$ is the probability associated with the discrete variable $x_i$.

Continuous probability distributions

The PDF provides the likelihood that the random variable assumes a particular range
of values for any continuous distribution function. By computing the area under the
probability density function curve over the range of values provides the probability
that the random variables will fall in that interval.

Two conditions should satisfy any PDF:

$$\begin{aligned}
&\text{(i)} \quad f(x) \geq 0 \;\; for\; all\; x \\
&\text{(ii)} \quad \int_{-\infty}^{\infty} f(x)dx = 1
\end{aligned} \tag{3.29}$$

We will describe some of the most common continuous probability distributions.

Continuous uniform probability distribution

A continuous uniform probability is one in which all the points have the same probability of occurring. For a continuous uniform random variable $x$ defined between two
extreme values $a$ and $b$, the uniform density function is described by the Eq. (3.30),

$$f(x) = \frac{1}{(b-a)} \quad where\; a \leq x \leq b \tag{3.30}$$

For values of $x$ outside this range, $x < a$ or $x > b$, $f(x) = 0$.

The uniform distribution is sometimes called a "rectangular" distribution or
random distribution. There are some cases in which the uniform probability distribution can be used to represent the uncertainty of a variable (assuming all values
are equally likely). However, the primary use of this distribution is for sampling in
another distribution during the Monte Carlo simulation, as we will discuss in Chap. 4
of this book.

The mean value (or expected value) and the standard deviation of the uniform
density function are calculated using the Eqs. (3.31) and (3.32),

$$\mu = \frac{a+b}{2} \tag{3.31}$$

$$\sigma = \sqrt{\frac{(b-a)^2}{12}} \tag{3.32}$$

Figure 3.5 shows the uniform probability distribution between 1 and 2 and the
corresponding cumulative distribution.

Microsoft Excel does not provide any function for the uniform distribution.

In the R language, the uniform density probability distribution and the uniform
cumulative distribution are obtained by the functions shown in Eqs. (3.33) and (3.34),

$$= dunif(x, y) \tag{3.33}$$

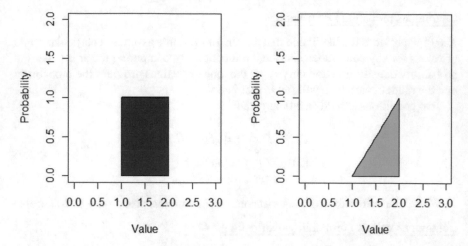

**Fig. 3.5** Uniform probability distribution

and

$$= punif(x, y) \tag{3.34}$$

Continuous Normal probability distribution
_____

The *Normal probability distribution* is probably the most important probability distribution function. The Normal distribution models many random variables which describe actual phenomena. It is also called *Gaussian distribution* in honour of the German mathematician, physicist and astronomer Carl Friedrich Gauss (1777–1855), who developed the mathematical formulation of this distribution (he also made outstanding contributions in several domains such as differential geometry, statistics, magnetism, optics, etc.).

A variable can be described by a Normal distribution when a continuous and symmetric variable is distributed following the shape of a bell curve. The Normal probability density function has the Eq. (3.35),

$$f(x) = \frac{1}{\sigma\sqrt{2\pi}} e^{\frac{-(x-\mu)^2}{2\sigma^2}} \qquad where -\infty < x < +\infty \tag{3.35}$$

In this equation, $\mu$ is the mean value, and $\sigma$ is the standard deviation of the distribution. Different selections of $\mu$ and $\sigma$ provide different Normal probability distribution functions; a Normal distribution can be entirely and univocally defined by the two parameters, $\mu$ and $\sigma$.

Figure 3.6 shows the Normal probability distribution for the parameters $\mu = 0$ and $\sigma = 2$; the left tail is highlighted in orange, and the right tail in blue. This function has a range between $(-\infty, \infty)$.

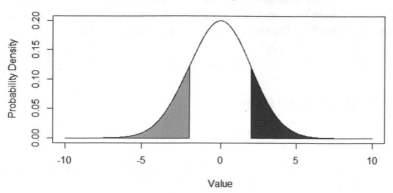

**Fig. 3.6** Normal probability distribution

**Table 3.1** Distribution of random variables

| Normal distribution | |
|---|---|
| Percentage of data points within interval, % | Interval |
| 68.26 | $\mu \pm \sigma$ |
| 95.44 | $\mu \pm 2\sigma$ |
| 99.74 | $\mu \pm 3\sigma$ |

Any normal random variable is distributed across its range following the proportions in Table 3.1.

These figures indicate that most data points are distributed within an interval of a few standard deviations from the mean value.

Because a Normal probability distribution is symmetric around the mean value, the mean, the median and the mode are equal.

Many phenomena are characterized by a normal or bell shape distribution; in these cases, the Normal distribution can be used to identify outliers in the data set.

When the parameters of the Normal probability distributions are $\mu = 0$ (the mean is zero) and $\sigma = 1$ (the standard deviation is one), this distribution is called a *standard Normal probability distribution*, and it satisfies the Eq. (3.36),

$$f(x) = \frac{1}{\sqrt{2\pi}} e^{-x^2/2/2} \tag{3.36}$$

In the subsurface domain core porosity, the percentages of certain chemicals or oxides in rocks are well described by a Normal distribution.

In the Normal distribution, the area below the PDF is an integral that does not exist in a closed formula but can be computed numerically. For a standard Normal probability distribution (a Normal distribution with $\mu = 0$ and $\sigma = 1$), the standard Normal distribution tables provide the cumulative probability for values less than the one chosen. The cumulative probability values for standard Normal probabilities can be transformed into any other Normal distribution with a different mean and standard deviation by using the transformation shown in Eq. (3.37),

$$z = \frac{x - \mu}{\sigma} \tag{3.37}$$

In Microsoft Excel, the function

$$= NORMDIST(x, mean, sd, TRUE) \tag{3.38}$$

provides the cumulative probability until value $x$, for a Normal probability distribution with mean "*mean*" and standard deviation "*sd*".

In the R language, the function

$$cum = pnorm(x, mean, sd) \tag{3.39}$$

provides the cumulative probability until value $x$, for a Normal probability distribution with mean "*mean*" and standard deviation "*sd*" (Fig. 3.7).

Continuous Exponential probability distribution

This probability distribution describes phenomena such as the time between arrivals and the distance between occurrences; an Exponential probability distribution is related to the discrete Poisson probability distribution. The equation describing the Exponential probability distribution is,

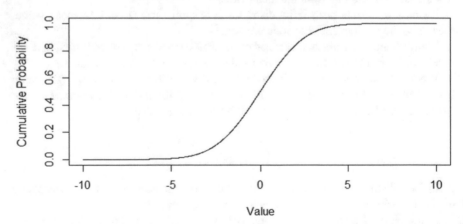

**Fig. 3.7** Cumulative normal probability distribution

$$f(x) = \frac{1}{\mu}e^{-x/\mu} \quad where \ x \geq 0, \mu > 0 \tag{3.40}$$

For the Exponential distribution, $\mu$ represents both the mean and the standard deviation. The cumulative distribution of the Exponential function is shown in Eq. (3.41) (Figs. 3.8 and 3.9),

$$p(x \leq x_0) = 1 - e^{-x_0/\mu} \tag{3.41}$$

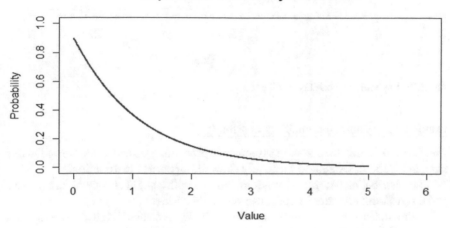

Fig. 3.8 Exponential probability distribution

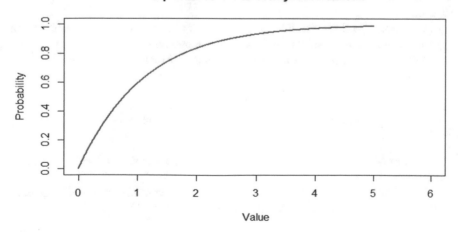

Fig. 3.9 Exponential cumulative probability distribution

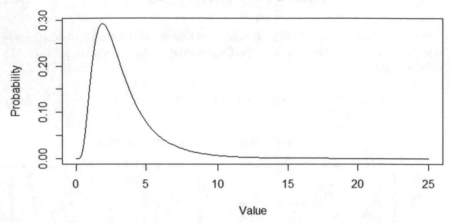

**Fig. 3.10** Lognormal probability distribution

Continuous Lognormal probability distribution

The Lognormal distribution is an asymmetric probability distribution, skewed to one side; in Fig. 3.10, we show a positively skewed Lognormal distribution.

This distribution describes a random variable with a slight chance of large values and a significant opportunity for lower numerical values.

In the subsurface reservoirs, several variables are distributed according to a Lognormal distribution, such as core permeability, oil recovery in a reservoir subject to the same reservoir mechanism and thicknesses of sedimentary beds.

In the same way, as in the Normal distribution, two parameters, the mean and the standard deviation are enough to completely and uniquely define the Lognormal distribution.

The range of values for the Lognormal distribution is $(0, \infty)$ because the logarithm function is undefined for 0 or negative numbers.

The mathematical equation for the Lognormal distribution is,

$$f(x) = \frac{e^{-\left(\frac{\left(\frac{\ln((x-\theta))}{m}\right)^2}{2\sigma^2}\right)}}{(x-\theta)\sigma\sqrt{2\pi}} \quad where \ x > \theta; m, \sigma > 0 \tag{3.42}$$

$\sigma$ is the standard deviation, $\theta$ is the location parameter, and $m$ is the scale parameter.

If $\theta = 0$ and $m = 1$, the distribution is called a *standard Lognormal distribution*. In that case, the Lognormal distribution is given by Eq. (3.43),

$$f(x) = \frac{e^{-\left(\frac{(\ln x)^2}{2\sigma^2}\right)}}{x\sigma\sqrt{2\pi}} \quad where \ x > 0; \sigma > 0 \tag{3.43}$$

In Microsoft Excel, the Lognormal distribution is obtained with the function,

$$= LogNorm.Dist(x, mean, sd, FALSE)$$

The Lognormal cumulative distribution can be described with the same function by changing FALSE to TRUE.

In the R language, the Lognormal density function is generated using the function shown in Eq. (3.44),

$$dlnorm(x, mean, sd) \qquad (3.44)$$

while the cumulative Lognormal distribution is generated using the function,

$$plnorm(x, mean, sd) \qquad (3.45)$$

Figure 3.11 shows the Lognormal cumulative distribution function.

Continuous triangular probability distribution

The triangular probability distribution is characterized by a triangular shape, symmetrical or skewed in either direction; the triangular distribution has a minimum and maximum values. When the mode coincides with one of the extreme values, the triangle has a right (90°) angle.

The triangle is completely specified given the minimum or low ($L$), maximum or high ($H$) and most likely ($M$) values. Thus, the mean and the standard deviation of a triangular distribution can be estimated using L, H and M as below:

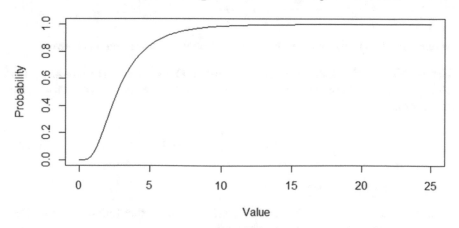

**Fig. 3.11**  Cumulative lognormal probability distribution

$$\mu = \frac{L + M + H}{3} \tag{3.46}$$

$$\sigma = \sqrt{\frac{(H - L)\left(H^2 - LH + L^2\right) - MH(H - M) - LM(M - L)}{18(H - L)}} \tag{3.47}$$

Discrete probability distribution

Any discrete random variable can take either a finite number of values (binomial probability distribution) or an infinite sequence of values (Poisson probability distribution); the latter category means that the maximum number of values is unknown.

The equivalent to the probability density function for a continuous variable is called the *Probability Mass Function* (PMF) in the case of a discrete variable; for a discrete value, we can assign a probability to any given value. The PMF $f(x)$ must always satisfy the following two requirements:

(i)    $0 \leq f(x) \leq 1$, the probability function is always higher than or equal to 0 and lower than or equal to 1.
(ii)   $\sum f(x) = 1$, the sum of all probabilities must be equal to 1.

Discrete uniform probability distribution

The uniform probability distribution assigns the same probability to all values inside the range where the function is defined; in other words, for random variables that follow a uniform probability distribution, all values of $x$ are equally likely. The discrete uniform probability function is,

$$f(x) = \frac{1}{n} \tag{3.48}$$

where $n$ is equal to the number of values that the random variable may have.

**Example 3.4: Uniform Probability Distribution**   One example of the uniform probability distribution is the toss of a fair die; the sample space is $S = \{1, 2, 3, 4, 5, 6\}$; in this example,

$$f(x) = \frac{1}{n} = \frac{1}{6} = 0.1667 \tag{3.49}$$

Binomial probability distribution

A binomial distribution is satisfied by a random variable that follows the so-called binomial experiment, which is described below:

(i)    The experiment consists of $n$ identical trials.
(ii)   In each trial, there are two possible outcomes, success and failure.

(iii)    From each trial, the probability of success p(s) is fixed.
(iv)    The trials are independent.
(v)     The outcome of the binomial distribution is the probability of $x$ successes in
        $n$ trials.

The binomial probability distribution function $f(x)$ is described by Eq. (3.50):

$$f(x) = p(x|n, p) = \binom{n}{x} p^x (1 - p)^{n-x} \quad \text{for} \quad x = 0, 1, 2, \ldots n \qquad (3.50)$$

where $p(x|n, p)$ is the probability of $x$ successes in $n$ trials in which $p$ is the probability of success.

The binomial distribution is widely used in quality control of manufacturing processes to estimate the probability of defective pieces; also, in oil exploration activities, this probability distribution can be used to calculate the probability of a given number of successful wells in an exploration campaign. The binomial probability equation is a particular case of a Bernoulli process. Only two outcomes can result from a trial, success or failure, dry hole or discovery, head or tail.

When in the binomial distribution $n$ becomes large (greater than 20), and $p$ becomes small (less than 0.05), it can be approximated by the Poisson distribution (as will be discussed later in this chapter), which is easier to solve than the binomial.

At the other extreme, the binomial distribution can be approximated by the Normal distribution when $n$ is large, and $p$ is not close to zero, with the approximation being better for $p$ close to 0.5. Figure 3.12 shows a binomial probability distribution.

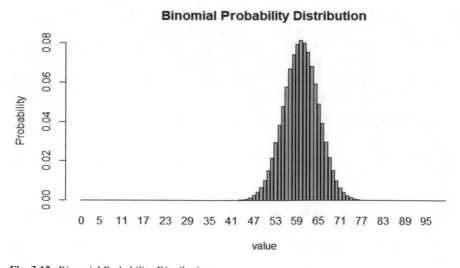

**Fig. 3.12** Binomial Probability Distribution

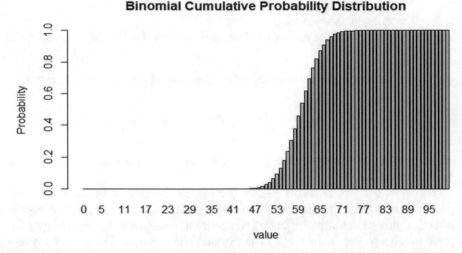

**Fig. 3.13** Binomial cumulative probability distribution

Figure 3.13 shows the cumulative probability distribution that corresponds to the binomial probability distribution shown in Fig. 3.12.

Poisson probability distribution

A random variable is described by the Poisson probability distribution when it is distributed across the range of values according to the Poisson experiments, which must satisfy three properties:

(i)     The probability of occurrence in any given unit of time, distance, area or volume is the same for all units of equal magnitude.
(ii)    The occurrence or non-occurrence in any unit of time, distance, area or volume is independent of the occurrence or non-occurrence in any other unit.
(iii)   The outcome of the Poisson distribution is the probability of $x$ occurrences of any given unit of time, distance, area or volume.

The Poisson probability distribution is given by Eq. (3.51):

$$f(x) = p(x|\mu) = \frac{\mu^x e^{-\mu}}{x!}, \quad for\ x = 0, 1, 2, \ldots . \tag{3.51}$$

where $f(x) = p(x|\mu)$ is the probability of $x$ occurring in a unit of time, distance, area or volume; $\mu$ is the mean number of occurrences in that unit of time, length, area or volume.

Figure 3.14 shows an example of the Poisson probability distribution for the case when $\mu = 7$.

Figure 3.15 shows the cumulative Poisson probability for the same case as in Fig. 3.14, where the mean is 7.

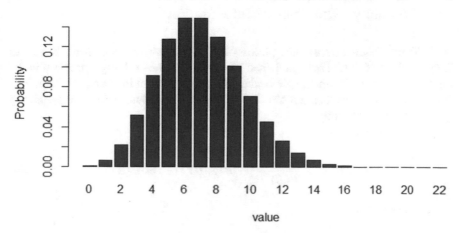

**Fig. 3.14**  Poisson probability distribution

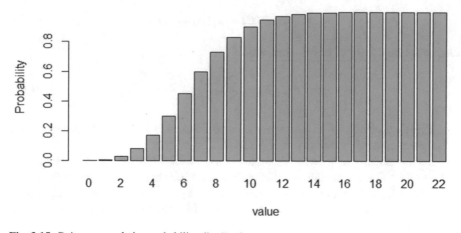

**Fig. 3.15**  Poisson cumulative probability distribution

In the R language, the Poisson mass function and the Poisson cumulative distribution are generated by the functions:

$$dpois(x, mean, FALSE) \,\&\, ppois(x, mean, TRUE) \tag{3.52}$$

In Microsoft Excel, the functions for the Poisson distributions are,

$$= Poisson.Dist(x, mean, FALSE) \,\&\, Poisson.Dist(x, mean, TRUE) \tag{3.53}$$

## 3.4  Formulation of the Bayes Theorem and Its Meaning—The Value of Information

In 1763 the English statistician, philosopher, and Presbyterian minister Reverend Thomas Bayes (1701–1761) published "An essay towards solving a problem in the doctrine of chances", where he introduced a theorem from then on took his name, Bayes' theorem. This theorem can be easily formulated considering the symmetry of conditional probability.

From Eq. (3.9),

$$P(A|B) = \frac{P(A \cap B)}{P(B)} \Rightarrow P(A|B)P(B) = P(A \cap B) \tag{3.54}$$

In the same way,

$$P(B|A) = \frac{P(A \cap B)}{P(A)} \Rightarrow P(B|A)P(A) = P(A \cap B) \tag{3.55}$$

Consequently,

$$P(A|B)P(B) = P(B|A)P(A) \tag{3.56}$$

or, in the final form of Bayes' theorem,

$$P(A|B) = \frac{P(B|A)P(A)}{P(B)} \tag{3.57}$$

This theorem allows us to assess the conditional probability of $A$ given $B$ based on the conditional probability of $B$ given $A$.

Let us assume that an event $E$, such as the possible Original Oil In Place (OOIP) of an oil field, the potential financial value of a project, etc., has a set of outcomes which the discrete set of values can describe $E_1, E_2, \ldots E_n$, each one corresponding to an event of nature; these values are all possible, each one with a probability of occurrence $P(E_i)$ (prior probabilities).

Let us assume that another event, an "information event" $A$, related with event $E$, has a value A; we will assume that, because $A$ is related with $E$, the conditioned probability of $E$ concerning $A$ is different from zero. However, $A$ is imperfect information, so it provides an indication of what could be the actual value of $E$, but uncertainty will remain, although probabilities associated with the outcomes $E_i$ will be updated in light of the new information.

The revised probabilities are calculated using Bayes' theorem, Eq. (3.57), with the result given by Eq. (3.58),

$$P(E_i|A) = \frac{P(A|E_i)P(E_i)}{\sum_{i=1}^{n} P(A|E_i)P(E_i)} \tag{3.58}$$

Bayes' theorem allows us to update the initial probabilities or likelihoods of occurrence, prior probabilities, for each possible *state of nature* $P(E_1)$, $P(E_2)$, ... $P(E_n)$ given the "information event" $A$; these conditional probabilities are $P(E_i|A)$ and these newly revised probabilities are called posterior probabilities.

Only one *state of nature* exists, but we do not know which one it is. For this reason, the description of nature is through a set of *states of nature*, each one with an associated probability of occurrence. Thus, Bayes' theorem's objective is to re-assess the probability of *states of nature* using the new information.

Some general comments regarding Bayes' theorem and its application for data acquisition:

(1)  Data acquisition can be considered as an iterative process when data is acquired in a sequence to reduce the uncertainty related to the true *state of nature*; this process allows us to make decisions with less uncertainty on the final outcome; in this approach, the aim is to reduce the uncertainty of the states of nature by increasing the probability of the most likely state and reducing the probability of the other ones.

(2)  Bayes' theorem can also be used to assess the value of information: for example, to determine how much money is recommended to spend on specific data acquisition actions which, by changing the probabilities of the states of nature, add value to the project.

These two approaches of data acquisition will be discussed in Chap. 7 of this book.

## 3.5  Sampling. Central Limit Theorem

*Random sampling* is the process used in the Monte Carlo simulation for selecting the observations in a sample. A *Simple Random Sample* (SRS) of size $n$ drawn from a finite population of si $N$ is one in which each sample of size $n$ has the same probability of being selected; the SRS is said to be unbiased. Several techniques can be used for sample selection without bias. If the sampling has a bias, the sample statistics do not represent the population from which the sample was taken.

For a sample of size $n$ selected from a population of size $N$, the total number of different samples is calculated as the combination of $N$ elements taken $n$ at a time, as described by Eq. (3.59):

$$C_n^N = \binom{N}{n} = \frac{N!}{n!(N-n)!} \tag{3.59}$$

and the probability of each sample is given by Eq. (3.60),

$$p(each\ SRS) = \frac{1}{C_n^N} \tag{3.60}$$

In the R language, we use the function,

$$choose(N, n) \tag{3.61}$$

For example, for samples of size ten taken from a population of size 90, there are a total of

$$C_{10}^{90} = \binom{90}{10} = \frac{90!}{10!(90 - 10)!} = 5.72e + 12 \tag{3.62}$$

So, there are $5.72e + 12$ different samples.

Sample statistics are also called *point estimators* because they are used to estimate the population parameter, e.g., $\bar{x}$ is a point estimator of the population mean $\mu$, $s$ is the point estimator of the population standard deviation $\sigma$ and $\bar{p}$ is the point estimator of $p$.

Apart from the SRS, other well-known sampling methods are: (i) systematic sampling, (ii) stratified random sampling, cluster sampling, (iv) convenience sampling, and (v) snowball sampling. These methods are briefly described below.

(i)   *Systematic sampling*: a probabilistic method; assume the size of the population is $N$, and the desired size sample is $n$; define the parameter $i = N/n$; randomly select one of the first $i$ elements and then select all the elements resulting from adding $i$ to the first element. For example, if $N = 50,000$ and $n = 100$, $i = N/n = {}^{50,000}/_{100} = 500$; selecting one random number from the first 500, e.g., 27, the remaining sample elements are 527, 1027, 1527....

(ii)  *Stratified random sampling*: a probabilistic method; first, the population is decomposed into mutually exclusive, collectively exhaustive subgroups or strata; using a random sampling method, such as SRS, items are drawn from each stratum. The strata are defined by stratification variables that represent a characteristic of interest. The intra-strata elements are as homogeneous as possible, and the intra-strata items are as heterogeneous as possible.

(iii) *Cluster sampling*: a probabilistic method; in this method, the population is decomposed into a mutually exclusive and collectively exhaustive cluster; a random method is used to select a sample of a cluster and finally include all the elements or a random sample of the elements of the cluster in the final sample. If all the elements of the selected clusters are included in the sample, the sampling is called one-stage cluster sampling. On the other hand, if a subsample of the cluster is drawn (probabilistically) from the cluster, this is called two-stage cluster sampling.

(iv)  *Convenience sampling* is a non-probabilistic method called accidental sampling; items are chosen according to what is convenient for the analyst to select. This sampling method is by nature inexpensive, easy and less time-consuming. However, the disadvantage is that the samples are not selected probabilistically, they are not representative of any population, so the results are doubtful. An example is surveying in shopping centres.

(v)   *Snowball sampling*: non-probabilistic; this is conducted in stages when first a set of individuals are selected; then, based on the outcome, another set is selected, and later other sets of individuals are selected as part of the sample.

## Central Limit Theorem

Similarly, as statistics are computed for one particular sample, statistics from a set of samples can also be calculated when several samples are taken. Each sample has different statistics: between two samples, their mean and standard deviations are probably different. So, for a set of samples, statistics can be computed. The probability distribution of the statistics is called a sampling distribution.

The expected value of $\bar{x}$, i.e., the mean of all possible sample means $\bar{x}$ is, by definition, equal to $\mu$, the population mean. This means, for example, that, in the case mentioned above, with a population size of 90 and sample size of 10, when all the $5.72e + 12$ samples are considered, the mean is equal to the population mean where the samples were collected. However, for a smaller number of samples, their mean, in general, is different from the population mean.

The standard deviation of $\bar{x}$ (standard deviation of the sample distribution) depends on the relation between the sample size and population size. There are two options:

(i)   If $n/N \leq 0.05$ then the standard deviation of the sample distribution, also called the standard error of the mean, is related to the standard deviation of the population and the sample size by the formula described by Eq. (3.63):

$$\sigma_x = \frac{\sigma}{\sqrt{n}} \tag{3.63}$$

(ii)  If $n/N > 0.05$ then the standard deviation of the mean distribution is given by Eq. (3.64):

$$\sigma_x = \sqrt{\frac{(N-n)}{(N-1)}} \frac{\sigma}{\sqrt{n}} \tag{3.64}$$

In the case mentioned above, where $N = 90$ and $n = 10$, $n/N = 0.11$, and by applying Eq. (3.64),

$$\sigma_x = \sqrt{\frac{80}{89}} \frac{\sigma}{\sqrt{10}} = 0.3\sigma \tag{3.65}$$

This equation represents the standard deviation of the sample distribution or the *standard error of the mean*.

Regarding the shape of the sample distribution, the *central limit theorem* states that, whatever is the population distribution function with mean $\mu$ and standard deviation $\sigma$ from which the samples were drawn, the sampling distribution of $\bar{x}$ approaches a Normal bell curve with mean $\mu$ and standard deviation $\sigma_x$, as described by Eqs. (3.63) and (3.64), as the number of samples approaches $C_n^N$.

More formally, the *Central Limit Theorem* (CLT) holds that, for samples randomly collected of size $n$ from any population distribution of mean $\mu$ and standard deviation $\sigma$, the sample mean distribution approximates a Normal distribution of mean $\mu$ and standard deviation $\frac{\sigma}{\sqrt{n}}$ as the sample size becomes larger, regardless of the population distribution shape.

This crucial theorem of probability theory was initially developed by the French mathematician Abraham de Moivre in 1733 and formally stated and named by the Hungarian mathematician George Polya in 1930.

Central Limit Theorem:

If $y_1, y_2, \ldots, y_n$ is a sequence of $n$ independent and identically distributed random variables with $E(y_i) = \mu$ and $V(y_i) = \sigma^2$ (both finite) and $x = y_1 + y_2 + \cdots + y_n$, then the limiting form of the distribution of de variable,

$$z_n = \frac{x - n\mu}{\sqrt{n\sigma^2}} \tag{3.66}$$

as $n \to \infty$ is the *standard Normal distribution*.

This theorem states that the sum of $n$ independent and identically distributed random variables is approximately normally distributed.

The CLT is a great tool to analyse the deviations of samples from population parameters.

For example, for an Exponential probability function of mean value 2.5, and taking 1,000 samples of size 40, sample A and 1,000 samples of size 4000, sample B.

Figure 3.16 shows at the left-hand side the more significant dispersion in the sample means for smaller simple size and opposite effect as the sample size increases.

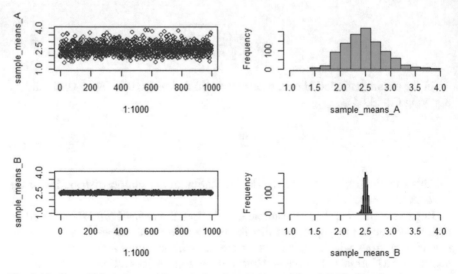

**Fig. 3.16** Central Limit theorem for sampling from Exponential probability functions

On the right-hand side, it is observable that the distribution is closer to the Normal distribution when the sample size increases.

Increasing the sample size generates samples with mean values closer to the population mean; this conclusion is summarized by the fact that the standard deviation in the first case is 0.3996874 and in the second case is 0.03975464. Besides, the mean value of the plot on the left is 2.535173, and the mean value of the plot on the right is 2.500751, which is much closer to the population value of 2.5.

## 3.6 Additional Probability Distributions

Now, we will review three important distributions: chi-square or $\chi^2$ distribution, t-student distribution and $F$ distribution, which are frequently used in hypothesis testing estimation.

### Chi-square distribution

Assume that $z_1, z_2, \ldots, z_k$ are random variables normally and independently distributed with mean 0 and variance 1, NID(0, 1); the random variable

$$x = z_1^2 + z_2^2 + \cdots + z_k^2 \tag{3.67}$$

follows a chi-square distribution with k degrees of freedom.

The density function of the chi-square distribution is described by Eq. (3.68):

$$f(x) = \frac{1}{2^{k/2}\Gamma(k/2)} x^{(k/2)-1} e^{-x/2} \qquad x > 0 \tag{3.68}$$

where $\Gamma\left(\frac{k}{2}\right)$ is the gamma function, defined by Eq. (3.69),

$$\Gamma(z) = \int_0^\infty t^{z-1} e^{-t} dt \quad z > 0 \tag{3.69}$$

$\chi$ distribution is skewed with mean and variance:

$$\mu = k \tag{3.70}$$

$$\sigma^2 = 2k \tag{3.71}$$

Figure 3.17 shows a set of $\chi$ distributions with different degrees of freedom.

The $\chi$ distribution is used to represent several meaningful expressions in statistics; for example, let us assume we have a random sample $y_1, y_2, \ldots, y_n$ extracted from a population described by $N(\mu, \sigma)$ distribution. It can be shown that,

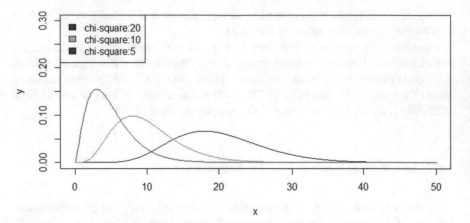

**Fig. 3.17** Chi-square distribution for 20, 10 and 5 degrees of freedom

$$\frac{SS}{\sigma^2} = \frac{\sum_{i=1}^{n}(y - \overline{y})}{\sigma^2} \sim \chi^2_{n-1} \tag{3.72}$$

which means that the variable $\frac{SS}{\sigma^2}$ is distributed as a chi-square distribution with $n-1$ degrees of freedom.

The term $SS$ will be used frequently in the ANOVA analysis in Chap. 8 of this book.

An example of a random number described by the chi-square distribution is the standard deviation,

$$S^2 = \frac{\sum_{i=1}^{n}(y - \overline{y})^2}{n-1} = \frac{SS}{n-1} \sim \frac{\sigma^2}{n-1}\chi^2_{n-1} \tag{3.73}$$

For example, the standard deviation of samples drawn from a population described by a Normal distribution $N(\mu, \sigma^2)$ is described by the chi-square distribution. So, the sampling distribution of the sample variance is a constant time chi-square distribution if the population is normally distributed.

t-student distribution

If $z$ is a standard normal random variable and $\chi^2_k$ is a chi-square distribution with k degrees of freedom, then $t_k$ follows a t-distribution or student's t distribution described by Eq. (3.74),

$$t_k = \frac{z}{\sqrt{\chi^2_k/k}} \tag{3.74}$$

which has $\mu = 0$ and $\sigma^2 = \frac{k}{(k-2)}$ for $k > 2$.

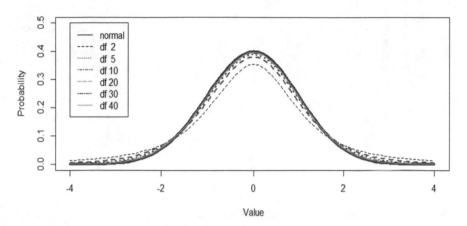

**Fig. 3.18** F distribution for different degrees of freedom: 2, 5, 10, 20, 30, 40 and standard Normal distribution

Figure 3.18 shows the t distribution for several degrees of freedom; as observed, the higher the number of degrees of freedom, the closer the distribution approaches the standard Normal distribution, and for $k = \infty$ the t distribution and standard Normal distribution are equal.

Let $y_1, y_2, \ldots, y_n$ be a random sample from a population of distribution $N(\mu, \sigma)$; the random variable described by Eq. (3.75) is distributed as t with n-1 degrees of freedom,

$$t = \frac{\bar{y} - \mu}{S/\sqrt{n}} \tag{3.75}$$

### F distribution

Let us assume we have two independent random variables having the chi-square distribution, $\chi_u^2$ and $\chi_v^2$, with $u$ and $v$ degrees of freedom, respectively; then the following ratio follows an $F$ distribution, given by Eq. (3.76),

$$\frac{\chi_u^2/u}{\chi_v^2/v} \sim F_{u,v} \tag{3.76}$$

with $u$ numerator degrees of freedom and $v$ denominator degrees of freedom. The probability distribution of a random variable $x$ that follows the $F$ distribution with numerator $u$ and denominator $v$ is given by Eq. (3.77),

$$h(x) = \frac{\Gamma\left(\frac{u+v}{2}\right)\left(\frac{u}{v}\right)^{u/2} x^{(u/2)-1}}{\Gamma\left(\frac{u}{x}\right)\Gamma\left(\frac{v}{2}\right)\left[\left(\frac{u}{v}\right)x + 1\right]^{u+v/2}} \qquad 0 < x < \infty \tag{3.77}$$

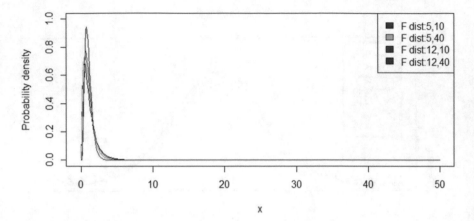

**Fig. 3.19**  F distribution with several degrees of freedom

Figure 3.19 shows examples of several $F$ distributions with different degrees of freedom.

In the R language, the function given in Eq. (3.78) provides the F distribution plotted in Fig. 3.19.

$$df(x, df1, df2) \tag{3.78}$$

where $x$ is a value and $df1, df2$ are the numerator and denominator degrees of freedom.

For a given cumulative probability, the corresponding value can be calculated by using,

$$qf(p, df1, df2) \tag{3.79}$$

For example, for 90 and 95%, the corresponding values are shown by Eqs. (3.80) and (3.81),

$$qf(.90, 20, 25) = 1.71752 \tag{3.80}$$

$$qf(.95, 20, 25) = 2.007471 \tag{3.81}$$

$F$ distribution is used frequently in statistics, e.g., for comparing two independent normal populations. For the random sample $y_{11}, y_{12}, \ldots, y_{1n_1}$ with $n_1$ observations from the first population and the random sample $y_{21}, y_{22}, \ldots, y_{2n_2}$ with $n_2$ observations and both populations with a common variance, $\sigma^2$,

$$\frac{S_1^2}{S_2^2} \sim F_{n_1-1,n_1-1} \tag{3.82}$$

where $S_1^2$ and $S_2^2$ are the sample variances.

## 3.7 Statistical Inference

The objective of statistical inference is to estimate population parameters using a sample from that population.

*Confidence interval*

In inference theory, *precision* is a probabilistic measure of the distance between a sample statistic and the population parameter that the statistics are supposed to estimate.

*The confidence interval* is a range of values that probably contain the population parameter of interest.

The *margin of error* is a value that is added to and subtracted from a point estimate to build the confidence interval of the population parameter of interest.

These two concepts, confidence interval and margin of error, can be joined in the Precision Statement (definition): given an estimate of $\bar{x}$, there is a probability (precision) $1 - \alpha$ that $\bar{x}$ has a sampling error equal to or less than $z_{\alpha/2}\sigma_{\bar{x}}$:

$$|\bar{x} - \mu| < z_{\alpha/2}\frac{\sigma}{\sqrt{n}} \tag{3.83}$$

where $\bar{x}$ is the sample mean, $n$ is the sample size, $\sigma$ is the population standard deviation, $1 - \alpha$ is the confidence coefficient and $z_{\alpha/2}$ is the value of the standard normal variable with area $\propto /2$ in any tail of the Normal probability curve.

Equation (3.83) can also be presented as per Eq. (3.84),

$$\bar{x} + z_{\alpha/2}\frac{\sigma}{\sqrt{n}} > \mu > \bar{x} - z_{\alpha/2}\frac{\sigma}{\sqrt{n}} \tag{3.84}$$

In the R language, we can use the function,

$$qnorm(x) \tag{3.85}$$

where x is the percentage required; this function provides the number that splits the range of values into two intervals: values of x for which the cumulative probability reaches x (the required probability) and values of x for which the cumulative probability reaches $1 - $ x.

When we look for a 95% precision, we mean that 2.5% of values are out of the range for the right side and the same for the left side—this is because of the symmetry of the Normal function.

Let us assume we are looking at the 95% interval probability; in that case $\alpha = 0.05$ and $\alpha/2 = 0.025$; it can be found in the standard Normal probability table or using R that $z_{\alpha/2} = z_{0.025} = 1.96$. Consequently,

$$\bar{x} - 1.96\frac{\sigma}{\sqrt{n}} > \mu > \bar{x} + 1.96\frac{\sigma}{\sqrt{n}} \tag{3.86}$$

This equation establishes the lower and higher limits of the population mean when 95% probability accuracy is required.

If we assume, e.g., that $\bar{x} = 20, \sigma = 3, n = 30$, there is a 95% probability that the population mean is located in the interval, and the error is 1.07, which means that the confidence is given by the Eq. (3.87),

$$21.07 > \mu > 18.92 \tag{3.87}$$

In the case where we are looking for a 90% interval probability, $\alpha = 0.10$ and $\alpha/2 = 0.05$, it can be found in the standard Normal probability table or using R that $z_{\alpha/2} = z_{0.05} = 1.64$.

$$20.90 > \mu > 19.10 \tag{3.88}$$

From Eqs. (3.87) and (3.88), we observe that as the confidence interval increases (or the confidence is lower), e.g., from 1.8 in Eq. (3.88) to 2.15 in Eq. (3.87), the confidence level (precision) increases, e.g., from 90 to 95%. Thus, the trade-off between confidence and precision is always present in the analysis. Therefore, the only remaining element that can influence the analysis is the sample size, which is the only factor that can improve the confidence and precision at the same time.

*Confidence level*: an interval estimate has associated a *confidence level* or probability value that the population parameter is inside the interval. For example, a 95% confidence interval means a 95% probability that the population parameters inside this interval, also called the referred interval, have a 95% confidence level.

*Confidence coefficient*: this is the confidence level expressed as a fraction, e.g., a 95% confidence level corresponds to $1 - \alpha = 0.95$.

In the previous discussion, it was assumed that the population's standard deviation is a known parameter used to estimate the unknown parameter, the mean of the population. However, in most cases, the standard deviation of the population is also unknown. In those cases, the previous analysis requires two changes: (i) instead of using the population standard deviation $\sigma$ we use the sample standard deviation $s$, and (ii) instead of using the standard Normal probability distribution, we use the t-distribution in such a way that the confidence interval is defined by Eq. (3.89):

$$\bar{x} \pm t_{\alpha/2,n-1} \frac{s}{\sqrt{n}} \tag{3.89}$$

The Eq. (3.89) can also be written as per Eq. (3.90),

$$|\mu - \bar{x}| < t_{\alpha/2,n-1} \frac{s}{\sqrt{n}} \tag{3.90}$$

The t-distribution is a set of probability distribution functions, similar to the standard Normal probability distribution. Each of the functions is differentiated by the parameter called the degree of freedom. By definition, $df = n - 1$, the sample size minus one.

The term $t_{\alpha/2,n-1}$ is the value of the variable $t$ that cuts the t-distribution into two parts, keeping an area of $\alpha/2$ in the upper tail of the curve, with n − 1 degrees of freedom.

In the R language, the function $qt(x, df)$ provides the number that splits the values into two parts: values of x. The cumulative probability reaches x and values of x, of which the cumulative probability reaches $1 - x$. In that sense, if the confidence level is 95%, then $\alpha/2 = 0.025$.

As discussed before, a trade-off between precision and confidence is always present in this type of analysis; however, the sample size is a variable that can be used to adjust the desired precision and confidence simultaneously.

Assume the planned *margin error* (confidence) is,

$$PME = \pm z_{\alpha/2} \frac{\sigma}{\sqrt{n}} \tag{3.91}$$

From Eq. (3.91), the sample size is estimated by using Eq. (3.92),

$$\sqrt{n} = \frac{(z_{\alpha/2})\sigma}{PME} \tag{3.92}$$

In the case when $\sigma$ can be estimated, Eq. (3.92) can be used to produce the desired PME with the corresponding precision level, $\alpha$.

For example, using Eq. (3.92) for a precision of 95% and a sample size of 30, the margin of error is 1.07; if we need to have the same precision but reduce the margin of error to 0.6, the number of samples results from Eq. (3.93),

$$n = \frac{(z_{\alpha/2})^2 \sigma^2}{(PME)^2} = 96 \tag{3.93}$$

This result means that for reducing the confidence from 1.07 to 0.6 while keeping the precision at 95%, the number of samples must be increased from 30 to 96.

*Hypothesis testing theory*

Hypothesis testing is a procedure to analyse data outcomes, and it is based on two elements:

(i)     The null hypothesis, and
(ii)    Alternative hypothesis.

The *null hypothesis*, $H_0$, is the condition tentatively assumed to be true unless compelling evidence throws doubt on it, in which case the null hypothesis is rejected; the *alternative hypothesis*, $H_a$ is the condition concluded to be true whenever the null hypothesis is rejected. If data shows that the null hypothesis does not hold, then it is rejected, and the alternative hypothesis is accepted.

For two samples, hypothesis testing can be formulated in the following manner:

$$H_0 : \mu_1 = \mu_2 \tag{3.94}$$

$$H_a : \mu_1 \neq \mu_2 \tag{3.95}$$

where $\mu_1$, $\mu_2$ are the mean values of the samples. The hypothesis testing procedure consists of developing a procedure to calculate the test statistics on the samples. The null hypothesis is then rejected depending on the value observed on the test statistics; the rejection or not depends on the rejection region or critical region.

*Test statistics* are the outcome of the hypothesis test that is used for rejecting or not rejecting $H_0$.

*The rejection region*, RR, is the range of values of the test statistics which lead to the rejection of $H_0$; RR should be defined before carrying out the hypothesis test.

When a hypothesis is formulated, $H_0$ represents the conservative or current case, with no change; it is assumed that $H_0$ will hold until evidence, data show otherwise and when that happens $H_0$ is rejected and $H_a$ is accepted.

Hypothesis testing is a well-established methodology for comparing two sampling sets to get conclusions about the populations they come from.

Given two samples, we can assume (hypothesize) that both samples come from a population with the same mean value, and the difference observed is due to the randomness associated with the particular sample. In that sense, this hypothesis assumes that the samples come from populations with equal means, which is the null hypothesis.

Hypothesis testing error types

Hypothesis testing can have two types of errors:

(i)     *type I error*: this happens when $H_0$ is rejected when it is true:

$$\alpha = (type\ I\ error) = P(reject\ H_0 | H_0\ is\ true) \tag{3.96}$$

(ii)   *type II error*: this happens when $H_0$ is not rejected when it is false; the probability of committing type I and type II errors are $\alpha$ and $\beta$, respectively. $\alpha$ is also called the level of significance.

$$\beta = P(type\ II\ erros) = P(fail\ to\ reject\ H_0|H_0\ is\ false) \qquad (3.97)$$

Often, for applying hypothesis testing, for the probability of error type I, $\alpha$, the significance level is fixed. Then the test procedure is designed for keeping a small value for the error type II, $\beta$.

**Example 3.5: Hypothesis Testing Using the Rejection Region** Let us consider the experiment of tossing a fair dice with six faces, numbered from 1 to 6, and consider two cases: draw the face with the number 6 on it and draw any other face but 6. Because the dice is assumed to be fair, each face has the same probability $\frac{1}{6}$ and the probability distribution of the outcomes can be represented by the binomial distribution: the probability that the dice will show the face numbered six is $\frac{1}{6}$ and the probability that the dice will show any other number is $\frac{5}{6}$. Figure 3.20 shows the probability distribution for the example in 25 trials (the dice is thrown 25 times).

At the top of each bar, the probability associated with each value is shown.

The RR acceptable for a test is a subjective criterion; here, we will define it as $RR_6$—the logic behind that number is that we have 25 trials and 0.16667 is the probability of getting 6 in one trial; so, for a fair dice, we can draw, until five times out of 25, the face with the number 6 (4.166 times). In this case, the level of significance is given by Eq. (3.98),

$$\alpha = p(x = 6) + p(x = 7) + \cdots + p(x = 25) = 0.23 \qquad (3.98)$$

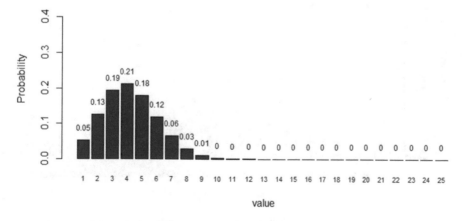

**Binomial Probability Distribution - Hypothesis Testing**

**Fig. 3.20** Binomial probability distribution for hypothesis testing

Using the R language, this value can also be estimated:

$$\alpha = 1 - pbinom(5, 25, 0.16667) = 0.23 \tag{3.99}$$

We can also relax the condition for rejection and reject the null hypothesis only if the face with the number 6 is drawn more than or equal to 6 times, which corresponds to $RR_7$. In this case,

$$\alpha = 0.11 \tag{3.100}$$

In this way, by reducing the rejection region, we reduce the level of significance, which drops from 23 to 11%. In practical terms, this means that in a dice trial, when the condition to accept that the dice is fair is relaxed, the level of significance is reduced, which translates to a probability reduction of the chance that error type I occurs by misinterpreting as an unfair dice one that is fair.

In the previous example, our concern was to assess the error type I, which is associated with the rejection of the null hypothesis when it is true; for that, we use the binomial probability of the null hypothesis case (1/6 for the face "6"). However, reducing $\alpha$ automatically increases $\beta$ (error type II), as shown in the next example.

**Example 3.6: Hypothesis Testing Using the Rejection Region II** Let us assume now that we have an unfair dice so that the probability associated with the face "6" appearing is 50%. In this example, we are interested in error type II, which assumes no rejection of the null hypothesis if it is false. For this case, the binomial probability that we will use is the one that applies to an unfair dice. The binomial distribution of this case is shown in Fig. 3.21.

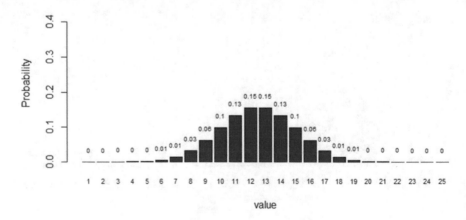

**Fig. 3.21** Binomial probability distribution for hypothesis testing—error type II

For this case, assuming a rejection region $RR_{13}$, the probability of the error type II occurring is

$$\beta = p(x = 0) + p(x = 1) + p(x = 2) + p(x = 3) + p(x = 4) + p(x = 5)+$$
$$p(x = 6) + p(x = 7) + p(x = 8) + p(x = 9) + p(x = 10)+$$
$$p(x = 11) + p(x = 12) = 0.654981$$

$$(3.101)$$

while when the rejection region is reduced, $RR_{15}$, the probability of error type II is,

$$\beta = p(x = 0) + p(x = 1) + p(x = 2) + p(x = 3) + p(x = 4) + p(x = 5)+$$
$$p(x = 6) + p(x = 7) + p(x = 8) + p(x = 9) + p(x = 10)+$$
$$p(x = 11) + p(x = 12) + p(x = 13) + p(x = 14) = 0.7878219$$

$$(3.102)$$

This equation means that when the rejection area is reduced, $\alpha$ is diminished, and $\beta$ is increased: the probability of error type I is diminished. Simultaneously, the probability of error type II is increased.

The RR approach provides a way to evaluate the probability, in a draw, of the data being rejected or not. However, in the hypothesis testing approach, we specify the value of the probability of committing a type I error or the significance level of the test.

### Hypothesis testing for the two-sample t-test

In the two-sample t-test, for the sample means of two populations with the same variance, the statistic for comparing these samples means is,

$$t_0 = \frac{\overline{y}_1 - \overline{y}_2}{S_p\sqrt{\frac{1}{n_1} + \frac{1}{n_2}}}$$

$$(3.103)$$

where $\overline{y}_1$ and $\overline{y}_2$ are the sample means, $n_1$ and $n_2$ are the sample sizes and $S_p$ is an estimate of the typical sample variance $\sigma_1 = \sigma_2$ :.

$$S_p^2 = \frac{(n_1 - 1)S_1^2 + (n_2 - 1)S_2^2}{n_1 + n_2 - 2}$$

$$(3.104)$$

where $S_1^2$ and $S_2^2$ are the individual sample variances.

The term $S_p\sqrt{\frac{1}{n_1} + \frac{1}{n_2}}$ is the standard error of the difference in means.

To decide whether the null hypothesis $H_0 : \mu_1 = \mu_2$ should be rejected, we evaluate the following inequality:

$$|t_0| > t_{\alpha/2, n_1+n_2-2}$$

$$(3.105)$$

**Table 3.2** Two-sample example

| Condition A | Condition B |
|---|---|
| 452 | 546 |
| 874 | 547 |
| 554 | 774 |
| 447 | 465 |
| 356 | 459 |
| 754 | 665 |
| 558 | 467 |
| 574 | 365 |
| 664 | 589 |
| 682 | 534 |
| 547 | 456 |
| 435 | 651 |
| 245 | 654 |
| 721 | 665 |

where $\alpha/2$ is half of the percentage point, or significance level, in the t-distribution with $n_1 + n_2 - 2$ degrees of freedom.

Equation (3.105) means that if $H_0$ is true, $t_0$ is distributed as $t_{n_1+n_2+2}$ and $100(1 - \alpha)$ per cent of the values of $t_0$ fall between $-t_{\alpha/2,n_1+n_2+2}$ and $t_{\alpha/2,n_1+n_2+2}$. If the sample produces a result outside those limits, it would be unusual if the null hypothesis is true, and that fact supports that $H_0$ should be rejected. The decision-maker decides the significance level.

**Example 3.7: Hypothesis Testing for the Two-Sample t-Test** Consider two samples sets, each containing 14 measurements of the same variables taken under two different conditions A and B, as shown in Table 3.2.

The summary of these two samples is included in Table 3.3.

The mean values observed are very close, 562 and 560; the data spread is higher in the sample set subject to condition A than the one subject to condition B. Therefore, many people may consider that the two sets are very similar. However, the hypothesis test can show if that is true or not.

The hypothesis test can be formulated in the following way:

$$H_0 : \mu_A = \mu_B \qquad (3.106)$$

**Table 3.3** Summary of two samples

|  | Condition A | Condition B |
|---|---|---|
| Mean | 562 | 560 |
| Standard deviation | 169 | 112 |
| Sample size | 14 | 14 |

$$H_A : \mu_A \neq \mu_B \tag{3.107}$$

If we choose $\alpha = 0.05$, we will reject the null hypothesis if the numerical value of the test statistics fulfils,

$$|t_0| > t_{0.025,26} = 2.056 \tag{3.108}$$

Using Eq. (3.104),

$$S_p^2 = \frac{(n_1 - 1)S_1^2 + (n_2 - 1)S_2^2}{n_1 + n_2 - 2} = 20488 \tag{3.109}$$

and, from Eq. (3.103),

$$t_0 = \frac{\bar{y}_1 - \bar{y}_2}{S_p\sqrt{\frac{1}{n_1} + \frac{1}{n_2}}} = 0.034328 \tag{3.110}$$

From these calculations, $t_0 < t_{0.025,26}$ and we fail to reject the null hypothesis based on the computed value of the test statistic.

**Example 3.8: Hypothesis Testing for the Two-Sample t-Test** In this example, we will use a different two-sample set, described in Table 3.4.

The summary of these two samples is included in Table 3.5.

For this example,

**Table 3.4** Two-sample example

| Condition C | Condition D |
|---|---|
| 531 | 420 |
| 721 | 547 |
| 422 | 704 |
| 651 | 362 |
| 422 | 359 |
| 801 | 365 |
| 623 | 461 |
| 311 | 211 |
| 632 | 201 |
| 913 | 312 |
| 892 | 429 |
| 729 | 592 |
| 199 | 322 |
| 833 | 824 |

**Table 3.5** Summary of two samples

|                    | Condition C | Condition D |
|--------------------|-------------|-------------|
| Mean               | 620         | 451         |
| Standard deviation | 218         | 180         |
| Sample size        | 14          | 14          |

$$t_0 = 2.241 > t_{0.025,26} = 2.056 \qquad\qquad (3.111)$$

which means that the null hypothesis is rejected, and the two samples are statistically different, $\mu_C \neq \mu_D$.

This analysis was done assuming $\alpha = 0.05$; however, if this condition is relaxed and we use $\alpha = 0.02$,

$$t_0 = 2.241 < t_{0.01,26} = 2.479 \qquad\qquad (3.112)$$

This result implies that by using this specified significance level, we fail to reject the null hypothesis.

Therefore, the null hypothesis is rejected at the 0.05 significance level but fails to be rejected at the 0.02 significance level.

The statistical analysis results are often obtained, as discussed above, using Eq. (3.105). However, controversy may arise because of the selected $\alpha$ value, the significance level: how far is the current value from the significance level to conclude that the null hypothesis is rejected. Can we choose another value and not a fixed value? Do other people agree with using the same significance level?

To answer these questions, the *p-value* approach is used to interpret the significance of the t-distribution.

*P-value*: this is an extremely popular measure of deviation from the null hypothesis. The p-value is the probability that the test statistics take a value that is at least as extreme as the observed value of the statistics when the null hypothesis is true. In other words, the p-value corresponds to the observed level of significance calculated in a sample result that is as unlikely as the observation if the null hypothesis is true. When $p < \alpha$ the null hypothesis is rejected, and when $p > \alpha$ the null hypothesis is not rejected. In most statistical studies $\alpha = 0.05$ and when $p < 0.05$, strong doubt about the validity of the null hypothesis arises; sometimes, we use $\alpha = 0.01$ or $\alpha = 0.1$. Using the p-value, we compare the *observed level of significance*, the p-value, with the *level of significance*, $\alpha$.

The confidence interval for the two-sample t-test

In the method of the two-sample t-test just described, we show the power of the hypothesis testing technique for concluding the closeness of the population means. However, we are often more interested in defining, for a given likelihood, the interval where the difference of the population means is located. In this case, we may know that the means of the two populations are different, and the objective is to define an interval and the associated probability that measure their differences.

Using Eq. (3.113),

$$|(\overline{x}_1 - \mu_1) - (\overline{x}_2 - \mu_2)| < t_{\alpha/2,n_1+n_2-2}S_p\sqrt{\frac{1}{n_1} + \frac{1}{n_2}} \qquad (3.113)$$

which can be arranged as

$$\overline{x}_1 - \overline{x}_2 - t_{\alpha/2,n_1+n_2-2}S_p\sqrt{\frac{1}{n_1} + \frac{1}{n_2}} \leq \mu_1 - \mu_2 \leq \overline{x}_1 - \overline{x}_2$$

$$+ \, t_{\alpha/2,n_1+n_2-2}S_p\sqrt{\frac{1}{n_1} + \frac{1}{n_2}} \qquad (3.114)$$

Example two-sample t-test with a confidence interval

Using the same data shown in Table 3.3 from Example 3.7, and assuming $\alpha = 0.05$,

$$8.35 \leq \mu_1 - \mu_2 \leq 329.65 \qquad (3.115)$$

This result means that the 95% confidence interval estimate on the difference in means is between 8.35 and 329.65; put another way,

$$\mu_1 - \mu_2 = 169 \pm 160.65 \qquad (3.116)$$

Equation (3.116) implies that the mean is 169, and the estimate's accuracy is $\pm 160.65$. The difference in means is never zero, so with a 5% level of significance, the means are not equal, so the null hypothesis does not hold.

## 3.8  Summary

Chapter three provides the reader with the required probability background and the tools required to understand and apply decision-making methods, especially the value of information and flexibility. First, the chapter describes three different meanings of probability; then, it summarises the rules associated with the operation of probability and shows a few examples. Section 3.3 elaborates on the relationship between population and sampling and how the former can be described using the latter. In the next section, we review the main probability distributions and the parameters used for describing the uncertainty in the inputs and outcomes in many decision problems; examples are provided of these distributions in Excel and R language. Further, we explain Bayes´ theorem, which is a crucial concept underpinning the value of information methodology: it allows for updating the probabilities associated with the *state of nature* when new data is acquired. The following two sections of the chapter discuss sampling methods and statistical inference principles, which are essential techniques for managing data uncertainty and experimentation analysis.

# Reference

Savage, L. (1954). *The foundations of statistics.* Wiley.

# Chapter 4
# Probabilistic Evaluation of Uncertainties: Monte Carlo Method

**Objective**

The Monte Carlo method is a technique used in assessing the impact of the uncertainties in the input variables on the output variables. Current uncertainties in the variables (reservoir permeability, porosity, etc.) impact the future performance of the system (for example, a well's oil rate) and, in general, the project valuation (oil price, material cost, etc.). Monte Carlo simulation is a procedure used to randomly account for uncertainties and generate probability distributions of the outcomes. It is used in some problems of value of information and value of flexibility.

## 4.1 Introduction to Monte Carlo Simulation

The Monte Carlo Method consists of using randomness to explore deterministic values. It can be traced to the eighteenth-century work of Georges Louis Leclerc. This French scientist was called Conde de Buffon (1707–1788), and, in 1777, he proposed the "Buffon's needle" method for estimating the value of $\pi$. A horizontal paper with a grid of lines separated by length $d$ and a needle with the same size as the interlines distance was considered. When the needle is thrown over the paper, it has a probability of $\frac{2}{\pi}$ to cross a line, or after tossing the needle N times, $\pi = \frac{2N}{X}$ where $X$ is the number of times the needle intersects a line. A more general case would be if, for a uniform grid of parallel lines of distance $d$ apart, the probability when a needle of length $l < d$ is dropped crosses a line is $P_{hit} = \frac{2l}{\pi d}$.

By experimentation, several people have tried to estimate the value of $\pi$ by actually throwing needles, the most remarkable case being Lazzarini in 1901. Mario Lazzarini was an Italian mathematician who made 3408 tosses and found $\pi$ to be:

$$\pi = \frac{355}{113} = 3.1415929 \tag{4.1}$$

© The Author(s), under exclusive license to Springer Nature Switzerland AG 2022
M. J. Vilela and G. F. Oluyemi, *Value of Information and Flexibility*,
Petroleum Engineering, https://doi.org/10.1007/978-3-030-86989-2_4

During the Manhattan Project (an American/British/Canadian research and development project carried out during World War II to produce nuclear weapons), in the 1940s, Stanislaw Ulam (Polish mathematician, 1909–1984) and John von Neumann (Hungarian American physicist, mathematician, and computer scientist, 1903–1957) developed the Monte Carlo method. That development was supported by the earlier building of ENIAC, the first electronic computer.

In 1946, while convalescing from an illness, Ulam studied the use of random sampling for estimating the chances that a Canfield solitaire laid out with 52 cards will come out successfully (Eckhardt, 1987).

The origin of Monte Carlo's name came from Ulam's colleague Nicholas Metropolis (1915–1999), a physicist, who suggested the name (Metropolis, 1987) because Ulam's uncle used to travel to Monte Carlo and gamble money borrowed to his family.

John von Neumann discussed the Monte Carlo simulation application for investigating radiation shielding and the distance that a neutron would likely travel through various materials with Ulan in 1946. They realised that the diffusion process occurring during a nuclear explosion could be modelled by tracking the path of a sample of individual particles. This approach was taken due to difficulties in solving this problem analytically, even though they had all the data, such as the average distance a neutron can travel in a substance before it collides with an atomic nucleus or how much energy the neutron was likely to emit after a collision. In the Monte Carlo simulation framework, each particle would be subject to random collisions and other interactions. Still, they could be followed on time and the results subject to statistical analysis. The use of Monte Carlo simulation was essential to the development of nuclear fission. During these investigations, von Neumann used pseudo-random numbers instead of genuinely random numbers to generate quick results for the project's progress. Later in 1948, Fermi, Metropolis, and Ulan obtained Schrodinger's equation's eigenvalues for neutron capture in cores, using Monte Carlo simulation (Hammersley & Handscomb, 1975).

The first paper published on Monte Carlo simulation was *The Monte Carlo Method* (Metropolis & Ulam, 1949), published in the *Journal of American Statistical Society.*

*Stochastic simulation* means the same thing as *Monte Carlo simulation,* and even though it provides a better description of what it is about, Monte Carlo is the name more widely used.

The Monte Carlo method is used to solve analytically intractable problems that would otherwise require time-consuming and/or costly experimentation. After the first physics applications, it has been further applied in many domains that involve chances and uncertainty supported by the wider availability of computers, starting from the 90s. Monte Carlo has been abundantly used in statistics, computer science, operational research, market sizing, customer service management, astrophysics, molecular modelling, semiconductor devices, traffic flow simulations, environmental sciences, optimisation problems, signal processing, etc. (Furness, 2011; Wright, 2019).

Monte Carlo simulation is a technique for estimating the impact of risk and uncertainty in financial, cost, production, and other *forecasting models.*

A forecast is a plan for the future. It is built on assumptions (return of a project, cost of material or resources, time to complete the several tasks, etc.) which are estimates of future performance; these estimates have uncertainties and risks because they refer to an unknown value in the future.

Typically, the input variables' ranges can be estimated based on previous experience or common sense, and, consequently, the outputs will also be within ranges.

The Monte Carlo method takes the range of the output values for assessing the likelihood of outcomes to occur, based on the ranges estimated for the input variables. Taha (2006) described Monte Carlo simulation as a modelling scheme that estimates stochastic or deterministic parameters based on random sampling.

The main objective of Monte Carlo simulation is to quantitatively characterise the uncertainty and variability in estimates of exposure or risk and to identify critical sources of variability and uncertainty, as well as quantifying the relative contribution of these sources to the overall variance and range of model results (EPA, 1997).

The Monte Carlo method can be used: (i) to generate random variables for measuring their behaviour, (ii) to estimate numerical quantities by repeated sampling, and (iii) to solve complicated optimisation problems using a randomised algorithm.

Monte Carlo simulations mean any simulation that uses a random number in the simulation algorithm in the term's broadest sense.

## 4.2  Simple Monte Carlo Applications

Before describing the Monte Carlo workflow details, we describe an application of this technique for numerical integration.

As we know, the irrational number $\pi$ (3.141592653….) is one of the most important mathematical constants. The name comes from the Greek word describing the periphery and perimeter of a circle; the origins of this unique number go back to the Greek mathematician and physicist Archimedes (287–212 BC). $\pi$ was studied independently by the mathematicians William Oughtred (British, XV century), William Jones (Welsh, XVI century, who start using the symbol we use today) and Leonard Euler (Swiss, XVI century), and it measures the relationship between the length of a circumference and its diameter.

The area of a circle is given by Eq. (4.2)

$$A_c = \pi r^2 \tag{4.2}$$

where $r$ is the radius of a circle.

On the other hand, the area of a square is given by Eq. (4.3),

$$A_s = 4r^2 \tag{4.3}$$

**Fig. 4.1** The relation
between the areas of a circle
and the corresponding square

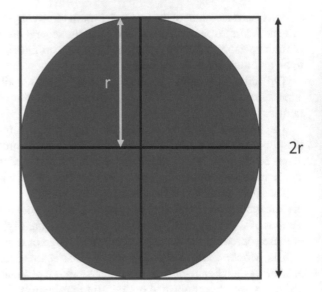

The relation between the area of a circle and the corresponding square is shown in Fig. 4.1.

The ratio of the circle area to the area of the square is given in Eq. (4.4).

$$\frac{A_c}{A_s} = \frac{\pi r^2}{4r^2} = \frac{\pi}{4} \tag{4.4}$$

One approach for estimating the value of $\pi$ is to generate a sample of $n$ pairs of uniform random numbers in the interval $[-r, +r]$. Each pair of random numbers define the $x$ and $y$ coordinates of a 2D point. Thus, t total number of points that fall into or over the circle and the total number of points generated are equal to $\pi$ divided by 4.

We developed a program in the R language for making this calculation. In Figs. 4.2, 4.3, 4.4, 4.5 and 4.6, we show four Monte Carlo simulation cases with an increasing number of points or iterations for getting approximations to the value of $\pi$. The points inside or in the circle's border are red, and the points outside the circle and inside the square are coloured in blue.

It is observed that, as the number of iterations increases, the approximation to the real value of $\pi$ improves.

Table 4.1 shows a summary of the estimations of the value of $\pi$ for several numbers of iterations.

The accuracy of the estimates measured as the percentage difference, concerning the actual value of $\pi = 3.1415926...$, increases as the number of iterations increases.

Based on the results shown in Table 4.1, the larger the number of iterations, the better the approximation of $\pi$. However, Table 4.1 only shows results that correspond to a single sample for each selected number of iterations. To better estimate, we can

**Monte Carlo Approximation of Pi = 3.204 with 1000 Iterations**

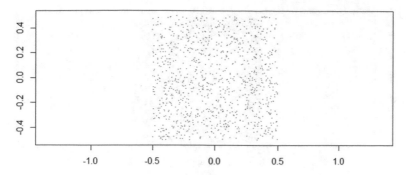

**Fig. 4.2** Monte Carlo approximation of Pi with 1,000 iterations

**Monte Carlo Approximation of Pi = 3.1608 with 10000 Iterations**

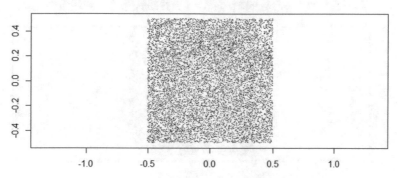

**Fig. 4.3** Monte Carlo approximation of Pi with 10,000 iterations

**Monte Carlo Approximation of Pi = 3.14444 with 100000 Iterations**

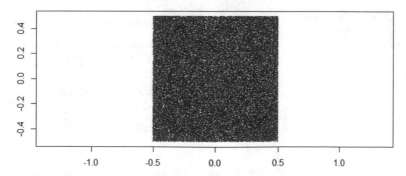

**Fig. 4.4** Monte Carlo approximation of Pi with 100,000 iterations

**Monte Carlo Approximation of Pi = 3.14242 with 1000000 Iterations**

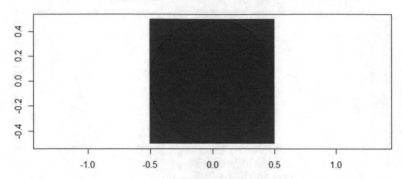

**Fig. 4.5**  Monte Carlo approximation of Pi with 1,000,000 iterations

**Monte Carlo Approximation of Pi = 3.1418736 with 10000000 Iterations**

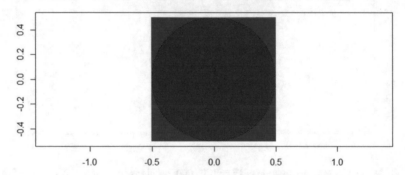

**Fig. 4.6**  Monte Carlo approximation of Pi with 10,000,000 iterations

| **Table 4.1**  Approximation of $\pi$ by increasing the number of iterations | Number of iterations | Approximate $\pi$ | Percentage difference, % |
|---|---|---|---|
| | 1,000 | 3.204000 | +1.986 |
| | 10,000 | 3.160800 | +0.611 |
| | 100,000 | 3.144440 | +0.090 |
| | 1,000,000 | 3.142420 | +0.026 |
| | 10,000,000 | 3.141874 | +0.009 |

generate several samples of each number of iterations and take the average. An algorithm in the R language generates the results shown in Table 4.2.

These estimated values of $\pi$ are obtained as the mean of the sample means.

The results from this example suggest that, for a desirable level of accuracy in the estimation of Monte Carlo simulation, a specific number of iterations is required.

**Table 4.2**  Approximation of π by increasing the number of iterations using 10,000 samples for several numbers of iterations

| Number of samples | Number of iterations | Approximate π | Percentage difference, % | Total difference |
|---|---|---|---|---|
| 10,000 | 1,000 | 3.140684 | +0.02892 | 0.00091 |
| 10,000 | 10,000 | 3.141653 | −0.00192 | −0.00006 |
| 10,000 | 100,000 | 3.141604 | −0.00036 | −0.00001 |
| 10,000 | 1,000,000 | 3.141566 | +0.00085 | 0.00003 |
| 10,000 | 10,000,000 | 3.141604 | −0.00036 | −0.00001 |

**Example 4.1: Monte Carlo Application to Project Management**  The Monte Carlo technique can be used to estimate the risk and uncertainties associated with project planning. For example, let us assume a construction project consisting of 4 sequential stages. The most likely estimated time is the period (in months) that the experts believe it will take to complete each stage. In addition, the maximum and minimum period required for each stage can also be estimated. Table 4.3 shows the time (in months) needed for completing each stage, using the best time estimates (most likely case), the minimum (optimistic) and the maximum (pessimistic) for completing each of the stages.

As shown in Table 4.3, the most likely scenario results in 41 months to complete the construction. However, it could take as little as 28 months or as long as 56 months. Thus, the most likely results may be a good outcome, but 56 months could be too long, impacting the company finances. For that reason, estimating the chance of finalising the construction at several periods is of great interest to the company. In such cases, the Monte Carlo simulation can obtain those estimates and capture the project's risk.

Monte Carlo simulation can be implemented using triangular functions for cases such as that above, where there are three-point estimates for each input variable. We would randomly generate values for each stage and then calculate the total time for the construction. In this example, the simulation was executed 50,000 times. Figure 4.7 shows the distribution results for the complete construction where each stage was estimated using Monte Carlo simulation for the corresponding minimum, medium and maximum values.

**Table 4.3**  Example 4.1: project management

| | Minimum number of months | Most likely number of months | Maximum number of months |
|---|---|---|---|
| Stage 1 | 2 | 5 | 8 |
| Stage 2 | 4 | 8 | 13 |
| Stage 3 | 10 | 12 | 16 |
| Stage 4 | 12 | 16 | 19 |
| Average | 28 | 41 | 56 |

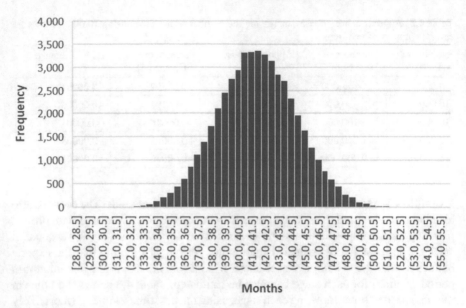

**Fig. 4.7** Monte Carlo simulation results using 50,000 simulations for the construction period in Example 4.1

The $S$ function describing the cumulative distribution of the time to complete the construction is shown in Fig. 4.8.

The quartiles of the results are:

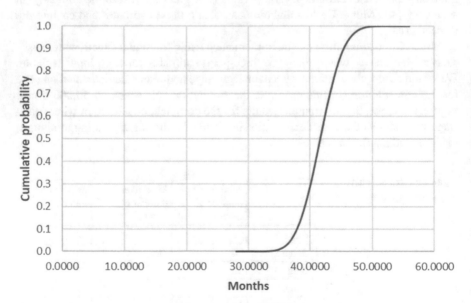

**Fig. 4.8** Cumulative probability result for the Monte Carlo simulation in Example 4.1

Q1 = 39.64
Q2 = 41.67
Q3 = 43.68
Q4 = 56.0

Data indicate that, even though the most likely case estimates that the construction can be completed in 41 months, there is just a 41.25% chance that the structure can be finalised in 41 months or less. Additionally, only a 29.04% chance the construction can be finished in less than 40 months, a target the construction company is pursuing. These are the project risks that can affect insurance, financing, permits, and hiring needs.

## 4.3  Number of Monte Carlo Simulations

One crucial question is how many simulations should be made to reach a specified level of accuracy? Unfortunately, this question, even though it is highly relevant, is usually underestimated.

Let us assume that $X_1, X_2, \ldots, X_m$ is a set of random samples of size $n$, generated from the population $X$. According to the Central Limit Theorem (CLT), if $n$ is sufficiently large, the mean of the random samples follows a Normal distribution with mean $\mu_{\bar{x}} = \mu$ and $\sigma_{\bar{x}} = \sigma/\sqrt{n}$.

The $100(1 - \alpha)\%$ confidence interval is estimated as per Eq. (4.5),

$$\bar{x} \pm z_{\alpha/2}\frac{\sigma}{\sqrt{n}} \tag{4.5}$$

which can also be written as shown in Eq. (4.6),

$$\left[-z_{\alpha/2}\frac{\sigma}{\sqrt{n}}, +z_{\alpha/2}\frac{\sigma}{\sqrt{n}}\right] \tag{4.6}$$

In this case, the margin of error (ME) is:

$$ME = z_{\alpha/2}\frac{\sigma}{\sqrt{n}} \tag{4.7}$$

If we define the preferred margin of error (PME) we are interested in reaching in the calculations, then $n$ should be calculated from Eq. (4.8),

$$n = \frac{\left(z_{\sigma/2}\right)^2\sigma^2}{PME^2} \tag{4.8}$$

When $\sigma$ is known, this equation can be used to directly estimate the number of samples $n$ that satisfy the desired precision (PME) and confidence ($\sigma$).

In the case that $\sigma$ is unknown, which is usually the situation in most real-world problems, we can use a pre-established figure for $\sigma$, obtained from a similar study from the past or a pilot study made with a smaller sample, that can be used as a reference to estimate $\sigma$. Then, based on this reference figure for $\sigma$, subsequent simulations can be run with the desired confidence level and the right precision.

In the case of sample proportions (proportions of individuals in the sample who have a particular characteristic), the preferred margin of error is calculated as per Eq. (4.9),

$$PME = z_{\alpha/2}\sqrt{\frac{\overline{p}(1-\overline{p})}{n}} \tag{4.9}$$

If Eq. (4.9) is re-arranged, we get Eq. (4.10):

$$n = \frac{\left(z_{\alpha/2}\right)^2 \overline{p}(1-\overline{p})}{PME^2} \tag{4.10}$$

Again, to evaluate Eq. (4.10), we need to know the value of $\overline{p}$ which is often unknown. A preliminary value from other studies can sometimes be taken as a starting point and then refined based on previous calculations. However, the most secure manner to estimate $n$ is by using Eq. (4.10); the number of $\overline{p}$ that maximises $n$ is $\overline{p} = 0.5$.

## 4.4 Monte Carlo Method Workflow

The Monte Carlo simulation is a simulation that depends on repeated random sampling and statistical analysis to compute the results. Monte Carlo uses mathematical models in natural sciences, social sciences, engineering, finance, and other domains to describe input variables' interactions that yield one or more outputs or objective functions.

In general, the Monte Carlo workflow consists of four steps:

1. *Create or identify the mathematical model of a system, process, or task under analysis.*
2. *Select the underlying probability distributions and their parameters that capture the input parameters' uncertainty.*
3. *Generate the random numbers from the distributions defined in the second step, which are specific values of the variables chosen randomly according to the distribution.*
4. *Analysis and decision-making by statistical analysis of the set of outcome values; this step will provide the* confidence required for making the decision.

The four steps are described below.

1. *Create or identify the mathematical model of a system, process, or task under analysis.* This is the deterministic model, which closely resembles the real scenario. This model has input variables and outcomes or objective functions or measures of the value of interest. The input variables can be separated into those known with certainty and those whose values are uncertain (when the analysis is being made) and will be modelled using random numbers. The model is initially evaluated using the most likely input parameters that produce the most likely outputs; this first assessment does not include any uncertainty consideration.

2. *Select the underlying probability distributions and their parameters that capture the input parameters' uncertainty.* This process is called distribution fitting. It consists of selecting the first set of probability distributions and their parameters that best fit each input parameter's historical data. Then, between the several distributions, the one closest to the information is chosen for sampling the data.

The first selection of probability distributions and parameters can be based on previous cases and expert knowledge, experience, or judgement; technical experts from different domains can assess the probability distributions to be used on various input variables. For example, in oil and gas exploration and production problems, a nearby field or other analogies can be used for defining unknown input parameters in the area of interest.

Even if the underlying probability distributions are unknown, all that is needed is that the expert draws a picture of the distribution of the input variables. With the existing algorithms, the probability distributions (shape and parameters) can be obtained. In most cases, different people in a team of experts provide insight into each input's best distribution. For example, geologists and petrophysicists should provide net thickness distributions in oil exploitation projects, reservoir engineers judge recovery factors, and drilling engineers provide input on well cost distributions. Typically, core porosity is reasonably described by a Normal probability distribution. Core permeability is well modelled by a Lognormal probability distribution, as is the case with reservoir thickness and oil recovery. Once the shape of the probability distribution is known, the missing information is the mean value and ranges, which in some cases can be obtained from analogues/nearby fields.

A technique called force-fit consists of estimating the low and high values of an outcome. For example, in an oil reservoir, a low and high value of the Original Oil In Place (OOIP), constructed by using pessimistic and optimistic values of the input variables and the assigned probabilistic percentiles to those values. A Normal or Lognormal distribution can then be 'forced' to fit those values (most often using a graphical representation of the fitting).

In general, the probability distribution shape is not as crucial as characterising its position (mean) and width (standard deviation).

When defining the probability distribution function of the variables, it is essential to know whether they have dependencies. When there are no dependencies, the values used for a variable $X$ has no effect on other variables $Y$ and $Z$. For example, on a particular trial, the value assigned to $X$ can be at the low end, and the value assigned to

$Y$ can be at the high end, but they do not impact the value of $Z$. However, when there are dependencies between the variables, these should be reflected in their values.

Dependencies between variables are often reported concerning oil production assessments: net pay thickness is related to well production rate (Darcy's equation), connate water saturation generally increases as porosity decreases, etc. These observations suggest that dependencies among variables should be included when defining the probability distributions used in a problem.

The distribution functions where the random numbers are generated for the simulations can be either continuous (for example, Uniform, Exponential, Gamma, Weibull, Normal, Lognormal) or discrete (for example, Bernoulli discrete uniform, binomial, geometric). In addition, these probability distributions can be parametrised with a few parameters that indicate location, scale, or shape, like the mean and standard deviation in the Normal distribution.

In some other situations, it is preferable to use observed data, also called *empirical distribution*, to specify the distribution from which random numbers will be picked for running the simulation. When using observed data, the data is either individual or group. Whatever the case, data is ordered from smallest to largest, representing the horizontal axis's values and the percentage of occurrence or frequency on the vertical axis. Points are interpolated linearly, creating a piecewise linear empirical distribution from which the random number can be extracted. The first selection criterion for defining the probability distribution is guided by the prior knowledge of the random variables' nature. For example, porosity is typically described with a Normal distribution. In many cases, distributions can be discarded based on theoretical considerations. One of the main disadvantages of using empirical distributions is that the random numbers generated are limited between the minimum and the maximum value of the data.

The comparison between data parameters and probability distribution parameters provides support for the selection of the probability distribution. For example, if the mean and the median of the data are close, this suggests a symmetric distribution. On the other hand, if the data's coefficient of variation is close to one, the Exponential distribution may be a good representation of the data. The skewness of a distribution $\nu$ is a measure of the symmetry: for symmetric distributions $\nu = 0$, such as in the Normal distribution, a distribution with $\nu > 0$ is skewed to the right, and if it is skewed to the left then $\nu > 0$. In this way, skewness can help with selecting the probability distribution function.

Histograms can also be used to select the probability distribution function; histograms are built by dividing the data into a given number of intervals of equal longitude. Each interval has an associated number that is the proportion of values included in the corresponding interval. The histogram, which is a piecewise constant plot, is then compared with different probability distribution functions. The one that resembles the histograms most is a good selection for representing the data. Thus, histograms can be used to uncover the probability distribution function applied to continuous and discrete variables. The main challenge when using this method is selecting the right number of intervals in the histograms. Suppose very short or very large intervals are taken. In that case, histograms will not identify the probability function (intervals that are too short generate erratic blocks and intervals that are too large generate big blocks). The most useful method to define the number of intervals in a histogram is by using the *Sturges rule,* which states that the number of intervals in the histograms $l$ should be given by Eq. (4.11).

$$l = (1 + log_2 n) = (1 + 3.322 log_{10} n) \tag{4.11}$$

We can consider the histogram to estimate the density or mass function of the probability distribution function.

Another indicator of the underlying probability distribution function is the quantile summary, which identifies whether the probability function is symmetric or skewed to the right or left. Given a random number and $0 < q < 1$, the quantile q of $F(x)$ is the number $x_q$ such that $F(x_q) = q$. Comparing the data for several quantiles, such as $x_{0.5}, x_{0.25}, x_{0.75}, x_{0.125}, x_{0.875}$ provides information on the shape and skew of the probability function.

Once the probability distribution function is identified by one of the methods described above (or a combination), the next step is to identify the distribution parameters. When the data is used to determine the distribution parameters, we say that *the parameters are estimated from the data.* An estimator is a numerical function of a given distribution, and the most common method for specifying the parameters is called the *maximum-likelihood estimators* (MLEs).

Suppose we have a discrete probability distribution for the data, which is described with one parameter, $\theta$; $p_\theta(x)$ denotes the probability mass function associated with the parameter $\theta$.

If we let the observed data be $x_1, x_2, \ldots, x_n$, we can define the likelihood function $L(\theta)$ by Eq. (4.12),

$$L(\theta) = p_\theta(x_1) p_\theta(x_2) \ldots p_\theta(x_n) \tag{4.12}$$

The likelihood function $L(\theta)$ is the joint probability mass function that gives the probability of occurrence of the observed data if $\theta$ is the value of the parameter that describes the distribution. If $\hat{\theta}$ is the value of $\theta$ that maximises $L(\theta)$, $L\left(\hat{\theta}\right) > L(\theta)$; thus, by maximising the likelihood function, the parameter describing the probability distribution function can be estimated.

MLE is asymptotically unbiased, meaning that the bias tends to zero as the number of samples increases to infinity. Also, it is asymptotically efficient; no unbiased estimator has a lower mean squared error than the MLE.

The MLE method evaluates a distribution parameter that fits the observed data from the application point of view. In that sense, given a probability distribution function, its parameter can be defined as a function of the data points. Table 4.4 shows a few probability functions and the corresponding MLE.

**Example 4.2: Derivation of MLE for an Exponential Distribution** In this example, we will derive the MLE for the Exponential distribution function, and the density function is given by Eq. (4.13),

**Table 4.4** Geometric, Lognormal, and Normal distribution probability functions and MLE

| Distribution function | Density or mass function | MLE |
|---|---|---|
| Geometric | $p(x) = \begin{cases} p(1-p)^x & x \in \{0, 1, \ldots\} \\ 0 & otherwise \end{cases}$ | $\hat{p} = \frac{1}{\bar{x}+1}$ |
| Lognormal | $f(x) =$ | $\hat{\mu} = \frac{\sum_{i=1}^{n} \ln(x_i)}{n}$ |
| | $\begin{cases} \frac{1}{x\sqrt{2\pi\sigma^2}} \exp\left(\frac{-(\ln(x)-\mu)^2}{2\sigma^2}\right) & if \ x > 0 \\ 0 & \\ & otherwise \end{cases}$ | $\hat{\sigma} = \left[\frac{\sum_{i=1}^{n}(\ln x_i - \hat{\mu})^2}{n}\right]^{\frac{1}{2}}$ |
| Normal | $f(x) = \frac{1}{\sqrt{2\pi\sigma^2}} \exp\left(-\frac{(x-\mu)^2}{(2\sigma)^2}\right)$ | $\hat{\mu} = \bar{x}(n)$ |
| | | $\hat{\sigma} = \left[\frac{n-1}{n} S^2\right]^{\frac{1}{2}}$ |
| Exponential | $f(x) = \begin{cases} \frac{1}{\beta} \exp(-x/\beta) & if \ x > 0 \\ 0 & otherwise \end{cases}$ | $\hat{\beta} = \bar{x}$ |

$$f_\beta(x) = \frac{1}{\beta} \exp(-x/\beta) \ for \ x \geq 0 \tag{4.13}$$

where the sub-index in $f$ is used to highlight the dependency of this function on the value of the parameter $\beta$.

When we have $n$ observed data points, the LME is given by Eq. (4.14),

$$L(\beta) = \left(\frac{1}{\beta} \exp\left(-\frac{x_1}{\beta}\right)\right)\left(\frac{1}{\beta} \exp\left(-\frac{x_2}{\beta}\right)\right) \ldots \left(\frac{1}{\beta} \exp\left(-\frac{x_n}{\beta}\right)\right) \tag{4.14}$$

which can be re-written in the following way:

$$L(\beta) = \left(\frac{1}{\beta}\right)^n \exp\left(-\frac{1}{\beta}\sum_{i=!}^{n} x_i\right) \tag{4.15}$$

The value of $\beta$ that maximises $L(\beta)$ is the same as the one that maximises the logarithm of $L(\beta)$ (logarithm is a strictly increasing function), which is easier to work; the value $\beta$ that maximises $L(\beta)$ is given by Eq. (4.16).

$$\ln(L(\beta)) = -n\ln(\beta) - \frac{1}{\beta}\sum_{i=1}^{n} x_i \tag{4.16}$$

To find the parameter value that maximises this function, we equalise the first derivative to zero and obtain Eq. (4.17),

$$0 = \frac{d\left(\ln(L(\beta))\right)}{d\beta} = \frac{-n}{\beta} + \frac{1}{\beta^2}\sum_{i=1}^{n} x_i \tag{4.17}$$

From Eq. (4.17), we get equality:

$$\beta = \sum_{i=1}^{n} \frac{x_i}{n} = \overline{x} \tag{4.18}$$

Up to this point, this value of β either minimises or maximises $L(\beta)$. To prove if it maximises $L(\beta)$, we need to investigate the second derivative, given by Eq. (4.19),

$$\frac{d^2 l}{d\beta^2} = \frac{n}{\beta^2} - \frac{2}{\beta^3}\sum_{i=1}^{n} x_i \tag{4.19}$$

This second derivative is negative because all $x_i$ in the Exponential distribution function are positive. Equation (4.19) ultimately proves that $\beta$, equal to the data mean, is the MLE for the Exponential distribution function.

When the distribution probability function depends on two parameters, such as Normal or Lognormal distributions, the MLE can be written as Eq. (4.20):

$$L(\theta, \beta) = p_{\theta,\beta}(x_1) p_{\theta,\beta}(x_2) \ldots p_{\theta,\beta}(x_n) \tag{4.20}$$

In this case, the procedure is similar to the one described for the Exponential function (Example 4.2) for one parameter but using two derivatives, one for $\theta$ and one for $\beta$, which generate two equations, each given the corresponding data observations parameters.

**Example 4.3: MLE for Exponential and Normal Distributions** Table 4.5 shows data set 1 and data set 2, for which we needed to find the probability distributions and their parameters.

The first data set seems to follow a Normal distribution, and the second one follows an Exponential distribution, as described by their histograms in Figs. 4.9 and 4.10.

For data set 1, which seems to fit a Normal distribution described by two parameters, the LME method estimates these parameters using the equations in Table 4.3, for example, Eqs. (4.21) and (4.22),

$$\hat{\mu} = \overline{x}(n) = 18.75 \tag{4.21}$$

The second parameter is:

$$\hat{\sigma} = \left[\frac{n-1}{n}S^2\right]^{\frac{1}{2}} = 2.02 \tag{4.22}$$

**Table 4.5** Data for Example
4.3: MLE for Exponential and
Normal distributions

| Data set 1 | Data set 1 | Data set 2 | Data set 2 |
|---|---|---|---|
| 19.1 | 17.6 | 1.3013334 | 1.0003433 |
| 20.2 | 18.9 | 0.0001421 | 0.0007390 |
| 17.4 | 18.5 | 0.0202735 | 0.0000371 |
| 16.0 | 16.4 | 0.0000542 | 0.0002190 |
| 16.2 | 17.6 | 4.2903003 | 0.1874050 |
| 17.3 | 18.4 | 0.0090367 | 0.0000278 |
| 23.5 | 16.3 | 0.0001500 | 2.0001717 |
| 21.4 | 16.9 | 6.3478920 | 1.0466399 |
| 20.4 | 20.1 | 0.0290023 | 1.0102590 |
| 18.7 | 23.1 | 0.5204801 | 0.0000035 |
| 17.3 | 19.9 | 1.8841119 | 4.0005603 |
| 19.1 | 19.1 | 0.7603321 | 1.1346088 |
| 19.7 | 22.1 | 0.4112945 | 0.0011428 |
| 19.1 | 18.5 | 0.0040205 | 0.0000378 |
| 18.9 | 19.4 | 0.0030502 | 14.0996400 |
| 14.3 | 18.6 | 0.0040758 | 8.3930481 |
| 18.6 | 17.9 | 0.0903280 | 0.0002852 |
| 17.4 | 17.5 | 0.1317022 | 0.0004856 |
| 20.8 | 22.3 | 10.8822400 | 14.9560120 |
| 18.1 | 16.7 | 0.0003520 | 0.0009305 |
| 20.1 | 16.5 | 0.0002331 | 0.0095114 |
| 16.1 | 15.2 | 2.0014229 | 0.0000671 |
| 20.9 | 16.8 | 0.0006432 | 2.9660042 |
| 22.1 | 20.1 | 0.0001281 | 0.0000348 |
| 20.4 | 19.9 | 0.0000020 | 0.1410669 |

For data set 2, which, according to Fig. 4.8, seems to fit an Exponential probability distribution, the parameter distribution is estimated by Eq. (4.23),

$$\hat{\beta} = \bar{x} = 1.59 \tag{4.23}$$

Once we have chosen a probability distribution function that models the data observed and estimated their parameters, the next step is to evaluate how good this probability function represents the data's underlying distribution. This first approach is heuristic, based on similarities observed in specific data plots and the distribution. Here we discuss some of those techniques.

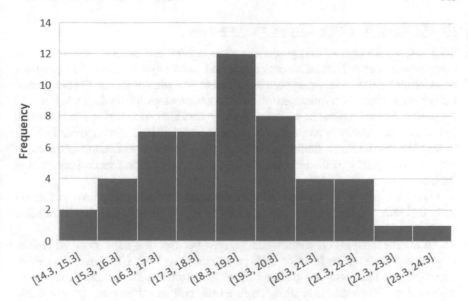

**Fig. 4.9**  Histogram of data set 1: Possible Normal distribution

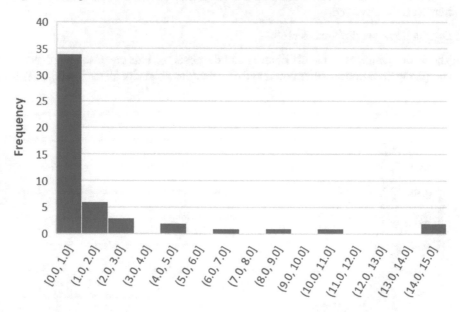

**Fig. 4.10**  Histogram of data set 2: Possible Exponential distribution

Density-histogram plots and frequency comparison

Both *density-histogram* and *frequency comparison* are similar techniques for comparing the data with the distribution function. In density-histogram, the histogram of the observed data is superimposed to the density function of the distribution probability function and observations are made as to whether they follow the same trend. In frequency comparison methods, the range of values is divided into several intervals or bins. The frequency or proportion of the observed data is plotted together with the probability distribution function in each interval. The visual inspection of the curve indicates the quality of the representativity of the distribution function concerning the data.

Figure 4.11 shows an example of the proportion comparison applied to data set 2. In this plot, we use the proportion so that the 200 points from the distribution function can be compared with the 50 points in the data set.

An excellent visual correlation can be seen between the data observed and an Exponential distribution function with parameter $\beta = 1.59$. Of course, the orange curve (Exponential probability distribution) is just one of the infinity of representations of the Exponential distribution function with the given parameter and 200 sample points. However, this can be checked for hundreds of selections, and similar trends can be observed.

Distribution-data differences plots

The quality of the fit of the distribution and the observed data can also be judged by comparing the number of data points and distributions points for all the intervals. This

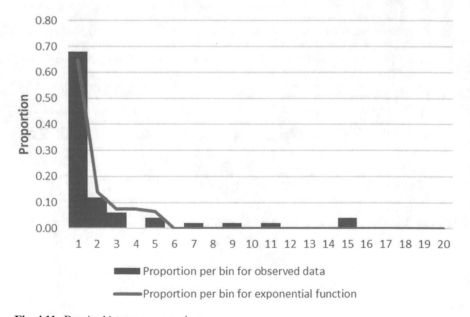

**Fig. 4.11**  Density-histogram comparison

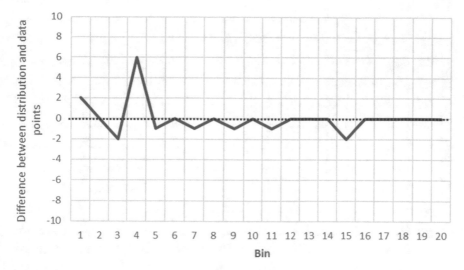

**Fig. 4.12** Differences plot between data set 2 and the Exponential probability distribution

comparison can be made using the cumulative distribution or the density function. Figure 4.12 shows the comparison using density function with 50 points to compare with the data observed.

In this realisation of the distribution function, it was noticed that the deviations are small. Several realisations of this distribution were tested, and similar results were obtained.

Figure 4.13 shows the cumulative comparison for data set 2 using the Exponential distribution function with $\beta = 1.59$ and 200 points.

Although there were a few differences for a few bins, there is a good match between the two curves.

The goodness of fit tests

There are several tests in this category; the *goodness of fit test* is a statistical hypothesis test to assess whether the observed data $x_1, x_2, \ldots, x_n$ are independent samples from a particular distribution function $F$.

As discussed in Chap. 3 of this book, hypothesis testing consists of proving that the null hypothesis is true or, otherwise, the alternative hypothesis is considered true. In this case, the null hypothesis is:

$$H_0 : the \ observed \ data \ x_1, x_2, \ldots, x_n \ IID \ are \ random \ variables$$
$$with \ distribution \ function \ F$$

The most common goodness of fit hypothesis test is the chi-square $\chi^2$ test, formulated by the English mathematician and biostatistician Karl Pearson (1857–1936).

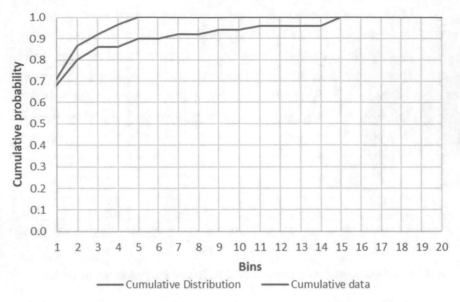

**Fig. 4.13** Cumulative comparison between data set 2 and the Exponential probability distribution

This test is based on the chi-square or $\chi^2$ distribution discussed in Chap. 3 of this book.

For the case when the fitted distribution is known, which means that the parameters are known, the $\chi^2$ test first divides the complete set of values in the sample into $k$ intervals, as in Eq. (4.24):

$$[b_0, b_1), [b_1, b_2), \ldots\ldots\ldots, [b_{k-1}, b_k) \tag{4.24}$$

The number of data points in each interval should then be counted, $N_j$ = number of sample points in the interval $[b_{j-1}, b_j)$.

For the fitted distribution, we estimate the probability associated with each of the previous intervals,

$$p_j = \sum_{a_{j-1} \leq x_j \leq a_j} p(x_i) \tag{4.25}$$

A similar equation is used when the random variables correspond to continuous distributions.

The statistic $\chi^2$ is defined by Eq. (4.26):

$$\chi^2 = \sum_{j=1}^{k} \frac{\left(N_j - np_j\right)^2}{np_j} \tag{4.26}$$

If $H_0$ is true, $\chi^2$ converges for large values of n to the $\chi^2$ distribution with k-1 degree of freedom (df), which is the same as the gamma distribution with parameters $[(k-1)/2, 2]$. This statement means that, if we want a confidence interval of $1-\alpha$, we will reject $H_0$ if we prove that $\chi^2 > \chi^2_{k-1,1-\alpha}$ where $\chi^2_{k-1,1-\alpha}$ is the value of the $1-\alpha$ critical point of the gamma distribution with parameters $[(k-1)/2, 2]$, i.e. the cumulative probability for the defined probability distribution that corresponds to the $1-\alpha$ confidence.

In the case when the parameters of the fitted distribution are unknown, but estimates are derived from the data by using MLEs, Chernoff and Lehmann (1954) showed that if $H_0$ is true, then when n is very large, the distribution function of $\chi^2$ converges to a distribution that lies between the chi distribution with $k-1$ and $k-m-1$ degree of freedom; this means that $\chi^2_{1-\alpha}$ satisfies Eq. (4.27).

$$\chi^2_{k-m-1,1-\alpha} \leq \chi^2_{1-\alpha} \leq \chi^2_{k-1,1-\alpha} \tag{4.27}$$

In this way, the value of $\chi^2_{1-\alpha}$ is only known between an interval. In general, most of the experts only prefer to reject $H_0$ in the case that the following inequity holds:

$$\chi^2 \geq \chi^2_{k-1,1-\alpha} \tag{4.28}$$

which is the conservative case.

One of the remaining problems for the chi-square test is the definition of the intervals' number and size. Unfortunately, there is no solution to this problem, and this issue is beyond the scope of this book.

3. *Generate the random numbers from the distributions defined in the second step, which are specific values of the variables chosen randomly according to the distribution*; for example, if the distribution that fits the historical data is Lognormal, numbers generated should fit a Lognormal distribution.

A set of random numbers is composed of a random number per each of the model's uncertain input parameters; each set of random numbers generates an outcome. Thus, many sets of random numbers are generated, and the model is evaluated many times (potentially into the millions), producing many results which can be further assessed and analysed. This step is the central part of the Monte Carlo simulation.

For the generation of random variables from any underlying probability distribution that corresponds to the uncertain input variables, we require the availability of a reliable independent and identically distributed (IID) $U(0, 1)$ random number generator, which are equally probable random numbers between 0 and 1. These random numbers are used in the simulation as an unbiased sampling method from the underlying probability distributions. A sequence of numbers is a *sequence of random numbers* if (i) there is no specific pattern in the order in which the numbers show in the sequence and (ii) each number in the sequence is equally likely to occur in the sequence.

There are different sources of random numbers:

(A)   *Noise in nature*: coming from a source such as atmospheric radio noise and lava lamps.
(B)   *Irrational numbers*: the sequence of digits to the right of the decimal point of irrational numbers, such as $\pi$, e, $\sqrt{2}$, etc.
(C)   *Physical experiment*: for getting random numbers between 1 and 6, we can toss a fair die with six faces as many times as the numbers of digits in the sequence.
(D)   *Reference book*: in the past, books were available according to random numbers (Rand Corporation published such a book in the 1950s).
(E)   *Random number generator*: this is currently the most used method to generate random numbers. It is a facility that most of the statistical applications, programmable tools and spreadsheets have available. However, most software applications do not have a random number but pseudo-random numbers; they are based on an equation that generates a uniformly distributed sequence of random numbers based on a seed. The equation is unknown to the user, which makes it impossible to know the sequence.

Random number generators should: (i) generate uniformly distributed numbers between 0 and 1 with no bias and a mean of 0.5, (ii) the sequence should not have self-correlation, which means that it is not possible to know the next value of a sequence even if we know the previous values, (iii) the series should not repeat (at least for very long series), and (iv) the user should have the possibility to set the seed (the same sequence will be obtained when the seeds are the same).

The uniformly generated random numbers are then used to create the random numbers from the probability distribution that best fits the data.

The most common method to generate random numbers that follow a specific probability distribution is the *Inverse Transformation Method*.

Inverse transformation method

For obtaining a sample of random numbers from a selected probability distribution function, we first need to know the inverse function of the associated cumulative distribution function. Using a uniform distribution function, we generate a sample of random numbers in the interval (0, 1). The random number is set equal to the numerically equivalent percentile of the cumulative distribution. These numbers are then used as inputs in the cumulative distribution function's inverse function to generate the sample of random numbers that follow the selected probability distribution function.

*Mathematically*: assume that $x$ is the continuous random variable that we want to generate, following a selected probability distribution function. Let us call $F$ the cumulative probability distribution function associated with $x$. $F^{-1}$ is the inverse function of $F$. For generating the sample of random numbers that follow the probability distribution function, we should follow three steps:

(1)   Generate a uniformly distributed random number: $y = Y(0, 1)$, where $Y$ is the uniform probability distribution.
(2)   Obtain $x = F^{-1}(y)$.

(3)   Repeat this process as many times as the size of the random number needed.

In this way, if 1,000 uniformly distributed random numbers are generated, 1,000 random numbers following the selected probability distribution function are obtained through the consecutive use of the inverse function $F^{-1}$.

**Example 4.4: Inverse Transformation Method for the Exponential Function.**
The Exponential density function is described by Eq. (4.29):

$$f(x) = \begin{cases} \dfrac{e^{-x/\alpha}}{\alpha} & if \ x \geq 0 \\ 0 & otherwise \end{cases} \qquad (4.29)$$

Figure 4.14 shows the Exponential density distribution function with media $\alpha = 20$.

Equation (4.30) describes the corresponding cumulative distribution function:

$$F(x) = \begin{cases} 1 - \exp(-x/\alpha) & if \ x \geq 0 \\ 0 & otherwise \end{cases} \qquad (4.30)$$

Figure 4.15 shows the cumulative distribution function of the Exponential function with media $\alpha = 20$.

For applying the inverse transform method, it can be proved that the inverse of the cumulative distribution function is given by Eq. (4.31),

$$F^{-1}(y) = -\alpha \ln(1 - y) \qquad (4.31)$$

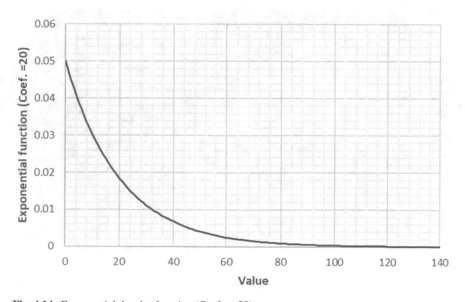

**Fig. 4.14**  Exponential density function (Coef. $= 20$)

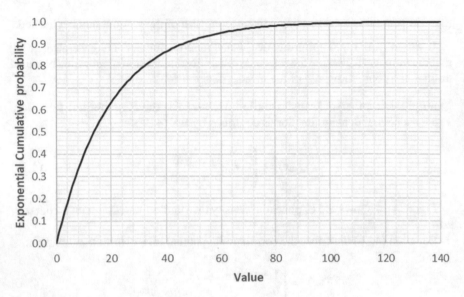

**Fig. 4.15** Cumulative Exponential distribution

In the inverse methods, a set of uniform random numbers are input in Eq. (4.30), and the results are a set of random numbers that follow the Exponential distribution.

From Fig. 4.16, it can be seen that taking random numbers uniformly distributed on the vertical axis expands different ranges in the horizontal axis. For example, if

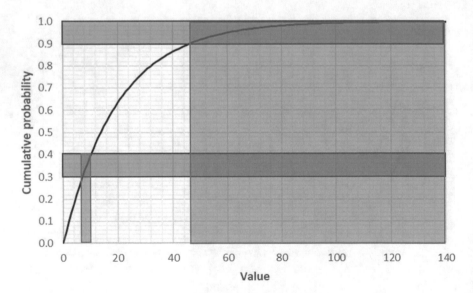

**Fig. 4.16** Inverse transformation method for the Exponential function

ten values are taken from the intervals (0.3–0.4) and (0.9–1.0), those values map onto the intervals (7–10) and (47–140) on the horizontal axis, respectively. These mappings are represented with blue and green rectangles in Fig. 4.16. The slope of the distribution function is much more prominent in the interval crossing (0.3–0.4) than in the interval crossing (0.9–1.0). This difference in the slopes is responsible for the difference in the ranges mapped in the horizontal axis. In this way, many generated variates fall in the interval (20–30), but not many in the interval (100–110); the density distribution function is the derivative (slope) of the cumulative distribution function. The density function has larger values, where the cumulative function has a larger slope. The inverse transform method spreads the uniformly distributed random numbers to follow the distribution function. The generated numbers are dense for ranges where the density function is high, and where the density function is low, the generated numbers are spread out. The inverse transformation deforms the uniform distribution to produce numbers that follow the desired probability distribution.

In the case of the discrete distribution function, $p(x_i)$ is the probability mass function for $x_i$ and it is written as per Eq. (4.32):

$$F(x) = \sum_{x_i < x} p(x_i) \tag{4.32}$$

For the inverse method, we should determine the smallest positive integral $I$ such that $y < F(x_I)$, and return $x = x_I$.

In this method for getting a random number from a selected probability distribution function, a random number coming from a uniform distribution between 0 and 1 is transformed, using the inverse of the desired probability density function (for continuous distributions) or probability mass functions (for discrete distributions), in a random value for the input distribution.

The main drawback of the inverse methods is that, in some cases, it becomes difficult to find a close form of the inverse cumulative distribution function for distribution. Alternatively, if CDF is challenging to get in a closed form, but the PDF can be quickly evaluated, iterative methods such as a bisection or Newton–Raphson can be used.

Another method used for generating random numbers from probability distributions is the acceptance-rejection method.

Acceptance-rejection method

An analytical form of the cumulative distribution function is often unknown or too complex to work with, and obtaining the inverse transform function is impractical. We assume that, for any value of x, the probability density function $f(x)$ can be computed. We also assume that there is a majoring function $t(x)$ such that $t(x) \geq f(x)$ for all x. In general, $t(x)$ is not a density function since, according to Eq. (4.33),

$$C = \int_{-\infty}^{\infty} t(x)dx \geq \int_{-\infty}^{\infty} f(x)dx = 1 \tag{4.33}$$

However, we can define the density function $r(x) = \frac{t(x)}{C}$.

The algorithm for the acceptance-rejection method is as follows:

1.   Generate Y from the distribution function $r(x)$.
2.   Generate U from $U(0, 1)$, independent from Y.
3.   If $U \leq \frac{f(y)}{t(x)}$ output $x$; *otherwise*, *return to step* 1.

The logic for the acceptance-rejection method is that for regions of $x$ where the difference between the distribution function $f(x)$ and the majoring function $t(x)$ are small, i.e., for high values of $f(x)$, more values of $x$ are taken, and the opposite is also true. In this way, the distribution points are picked so that the distribution of points selected follows the distribution $f(x)$.

**Example 4.5: Acceptance-Rejection Method for the Beta Function**  For the Beta (4, 3) function, presented in Eq. (4.34),

$$f(x) = \begin{cases} 60x^3(1-x)^2 \; if \; 0 \leq x \leq 1 \\ 0 \qquad\qquad\qquad otherwise \end{cases} \tag{4.34}$$

The C term in Eq. (4.33) is the maximum value of $f(x)$. Using derivatives, the value that maximises the Beta function is 0.6, and, at that point, the value of $f(x)$ is 2.0736.

Figure 4.17 shows the set of 100 values chosen uniformly and represented by the 'red cross' at value 2.00736 on the horizontal. These numbers are one realisation of 100 numbers, uniformly distributed. Figure 4.17 shows the randomly selected numbers represented by a purple cross at the bottom (x = 0). It is observed that the numbers are chosen in such a way that when the differences between the functions

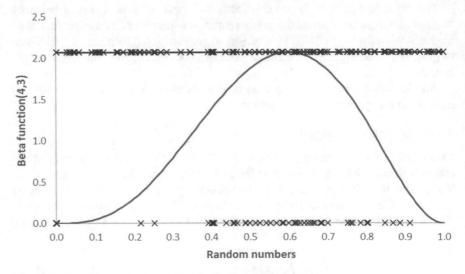

**Fig. 4.17**  Acceptance-rejection method for the Beta (4, 3) function

$f(x)$ and $t(x)$ are small, many numbers are selected. However, where this difference is significant, a small fraction of points are chosen.

If it is impossible to obtain an underlying probability distribution for an input variable due to the original distribution or data scarcity's complicated shape, the bootstrapped Monte Carlo simulation method, based on historical data, can be used. The data set (historical data) is repeatedly sampled by replacement, using a uniform random number generator to generate random integer numbers among an array's indices. Each element of the array is one member of the data set.

When there is more than one uncertain input variable in the mathematical equation describing the system, each variable should be sampled independently from the corresponding distributions. For $n$ variables, $n$ independent random numbers should be generated. In this way, different random numbers are used for each variable within a trial; this condition is necessary to avoid creating unrealistic correlations between the variables.

4.   *Analysis and decision-making by statistical analysis of the set of outcome values; this step will provide the* confidence required for making the decision.

We frequently need to make forecasts: return (benefit) of a portfolio of projects, cost of a construction project, future production of an oil field, time to complete a task, etc. We make forecasts to decide our actions today by assessing what will happen afterwards. However, the forecast is a projection of future outcomes and, consequently, has uncertainties. For example, we cannot know at present, with certainty, the future cost of an asset or the future production of an oil well. All we can do today is estimate expected (future) values and these uncertain.

When we have a deterministic model, a set of input variables are tied up by mathematical expressions in such a way as to produce one or more outcomes. For example, in a deterministic evaluation, the inputs' medium value is used to generate the medium value of the outputs. This case is called the base case or most likely case since these input variables' values are their most likely values. However, values are not typically known with certainty, and, in most cases, we can estimate, based on previous knowledge, expert opinion, partial information, etc., the ranges for the input variables. These are minimum, maximum and medium values usually called 'worst', 'best' and 'base' cases. These figures could be the minimum time to construct a building or the range of possible oil production rates of an oil well, etc. Using a range of values for the input variables will generate a range of values for the system outputs, making a difference with the standard or deterministic forecast where input values are provided with certainty, and the output values are obtained with confidence.

In the Monte Carlo simulation, the range of output values is associated with the probability of occurrence or how likely each particular outcome will occur. Monte Carlo simulation performs risk analysis by building models of possible results and replacing input variables with random values following a defined probability distribution function within the ranges estimated for the variables. The model or system is evaluated for this set of values, and an outcome is assessed. This process is repeated hundreds or thousands of times using different randomly selected values. Finally, these outcome values are used to estimate the average value of the system.

**Example 4.6: Application of Monte Carlo Simulation. Original Oil in Place** A Monte Carlo evaluation will be described to assess the uncertainties associated with the OOIP in an oil field.

1.  The first step for implementing the Monte Carlo method requires a known mathematical equation describing the system or process under study. The OOIP or volumetric estimation in a reservoir is estimated using Eq. (4.35),

$$OOIP = \frac{7758 \times \varnothing \times (1 - S_w) \times A \times h}{B_{oi}} \qquad (4.35)$$

where $\varnothing$ is the average porosity, $S_w$ is the average water saturation, $A$ is the area, and $h$ is the thickness of the reservoir, $B_{oi}$ is the oil formation volume factor and 7,758 is the conversion factor from acre-ft to barrels of oil (Bbls).

In this example, we will use the deterministic values for the variables, as shown in Table 4.6.

The deterministic OOIP = 902.1 MM STB.

STB means standard barrels; this figure is measured at surface conditions.

In this example, regarding the assessment of the OOIP, the parameters in Table 4.6 are the variables that were correctly applied in Eq. (4.35) to generate the value of interest, OOIP. However, there are other mathematical expressions in other applications in which several input variables are mathematically manipulated to yield an outcome that represents the value of interest.

2.  The second step is the identification of the distribution probability of the uncertain variables.

Based on expert knowledge, the probability distribution functions and their parameters indicated in Table 4.7 can be used.

3.  The third step in the Monte Carlo simulation consists of generating an unbiased set of random numbers for each distribution and the parameters shown in Table 4.6. In this example, we use the R language to create a uniform random number. Using the inverse method, we generate a set of random numbers for each distribution. For each member of the set, OOIP is calculated, as shown in Fig. 4.18 for 1,000,000 iterations.

4.  The fourth step is the analysis of outcomes. In this example, values were distributed in a histogram of 10 bars. The distribution shows that some values of OOIP are exceedingly small compared with the deterministic value, while

**Table 4.6** Deterministic values for the parameters in OOIP

| Parameter | Most likely value |
|---|---|
| Reservoir area (acres) | 10,000 |
| Reservoir thickness (feet) | 100 |
| Porosity (decimal) | 0.20 |
| Water saturation (decimal) | 0.25 |
| Oil formation volume factor (Bbl/STB) | 1.29 |

**Table 4.7**  Probability distribution functions for the input parameters

| Parameter | Probability distribution | Mean | Standard deviation | Minimum | Medium | Maximum |
|---|---|---|---|---|---|---|
| Reservoir area | Normal | 10,000 | 900 | | | |
| Reservoir thickness | Normal | 100 | 2 | | | |
| Porosity | Normal | 0.20 | 0.05 | | | |
| Water saturation | Triangular | | | 0.00 | 0.25 | 0.50 |
| Oil formation volume factor | Triangular | | | 1.05 | 1.29 | 1.45 |

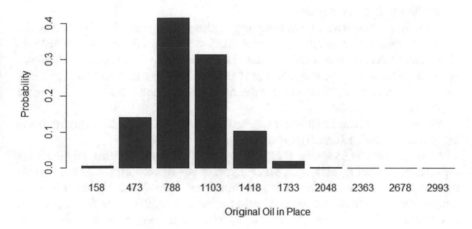

**Fig. 4.18**  Original oil in place distribution

others are up to three times larger. However, these extreme cases are doubtful, with a probability of occurrence less than 1%. Most of the results are between 700- and 1,100-MM STB.

Monte Carlo simulation allows the user to capture the risk and uncertainty of an objective variable or outcome by a range and distribution of the possible values obtained by considering the input variables' complete possibilities.

## 4.5   Another Sampling Technique: Latin Hypercube Sampling

Monte Carlos sampling is not the only way to sample numbers from a cumulative distribution of random variables. For example, Latin hypercube sampling was invented in the late 1970s and was used in the oil business in the 1980s. Latin hyper-cube sampling is an improved sampling method compared with the Monte Carlo method.

The random number generator's values are used for sampling points in the input variables' underlying cumulative probability distribution when using Monte Carlo.

Figure 4.19 shows 50 random numbers generated using Microsoft Excel. We can see that some values appear in clusters while there are sections in the interval (0, 1) that show gaps. This process makes the sampling in the underlying probabilities not entirely uniform. There is no reason for the clustering, and it is just a consequence of the randomness of random numbers.

This clustering effect tends to disappear when the number of simulations increases. However, this could be a limitation in the method, mainly when it is time-consuming to run the simulation many times.

There is an alternative way of sampling, called uniform sampling. To obtain 100 equally probable numbers in the interval (0.1), we need to divide the interval into 100 smaller intervals, each one longitude 0.01. The same procedure can be followed for each variable. The problem is that if the equation has four variables, we will need $100^4$ evaluations, which is a considerable number, and it is almost impossible to perform the simulation.

However, an alternative method sits halfway between Monte Carlo and uniform sampling; it is called *Latin Hypercube Sampling*.

Latin Hypercube Sampling (LHS) is a method of sampling that partitions the cumulative probability axis into a set of layers of equal probability. Assume that we divided the total interval into ten layers, numbered from 0 to 9. For each interval, we generate a random number and define a random number. The first digit after the decimal point is the number of the layer; the remaining digits are the random number generated for the layer in question. Proceeding in the same manner with the ten layers, each one will have a random number that can be used for sampling within the cumulative probability function of the input variable. This method forces one cumulative probability sampling point into each layer every ten trials of the simulation. A similar procedure should be followed for each input variable.

To generate the LHS, we just need to divide the complete interval into $n$ bins or intervals. Between the intervals $\frac{i-1}{n}$ and $\frac{i}{n}$ the random number, according to LHS, is given by Eq. (4.36):

$$y(i) = \left( \frac{i}{n} - \frac{i-1}{n} \right) Rand(*) + \frac{i-1}{n} = \frac{(Rand(*) + i - 1)}{n} \qquad (4.36)$$

**Fig. 4.19**  Clustering in the Monte Carlo sampling

Rand(*) is the random number generator that generates an equally probable random number in the interval (0, 1).

LHS has much less noise than Monte Carlo, which secures a quick convergence. Increasing $n$ makes convergence much faster. The number of trials should be a multiple of the number of layers $n$, which ensures that the sampling is made uniformly.

## 4.6   Advantages and Disadvantages of Monte Carlo Simulation

Monte Carlo simulation advantages

*Results are probabilistic*: the result of the simulation shows the possible outcomes and their associated likelihood.

*Graphical representation*: straightforward plots of the different outcomes and their chances of occurrence.

*Simplicity*: it is simple and straightforward as long as the convergence can be guaranteed.

*Solutions*: it provides approximated solutions to many mathematical problems.

*Correlation of inputs*: within the Monte Carlo framework, it is possible to model interdependencies between input variables as they occur in real-world problems.

*Scenario analysis*: Monte Carlo allows for the measurement of the impact that the input variables have on the system outcome.

*Sensitivity analysis* allows identifying the input variables that have the most considerable impact on the interesting outcomes.

Monte Carlo simulation limitations

*Equilibrium*: it is not appropriate to study problems that are not in equilibrium; it was developed for equilibrium systems.

*Many samples*: need to generate many samples and get statistics; one single sample is not enough.

*Accuracy*: the results of the Monte Carlo simulation are only an approximation to the true values.

*Time-consuming*: to reach a desirable result, many simulations may be required, which can be time-consuming, depending on the study.

## 4.7   Scope of Monte Carlo Simulation Applications

The Monte Carlo method has a broad domain of applications, as far as the input variables can be described with probabilities using random numbers. In the oil and gas industry, it has been used to estimate OOIP, recoverable reserves, Net Present

Value, and other issues where a mathematical equation describing the quantity of interest can be formulated. Kosova et al. (2015) discussed an example applying Monte Carlo simulation for estimating the reserve estimates of oil in the Kucova field (Ka-1 sector) in Albania. The Monte Carlo simulation application does not limit the number of input variables, although the larger the number, the more time is consumed in the calculations. A similar case was developed by Kok et al. (2006), who applied Monte Carlo to obtaining hydrocarbon reserves for an oil field located in Turkey.

Monte Carlo has been used to model biochemical processes (Tenekedjiev et al., 2011), inventory management, portfolio optimisation, project selection, reservation management, and benefits management (Evans & Jones, 2009). Furness (2011) illustrated the use of the Monte Carlo technique in three case studies from marketing and Customer Relationship Management (purchasing behaviour of the customer, cost centre resource scheduling, evaluation of customer analysis software). Platon and Constantinescu (2014) applied Monte Carlo analysis to perform a risk assessment for environmental projects by identifying and estimating risk and considering measures to mitigate their negative impacts (as required by the European Union for providing funds to support project selection). Song et al. (2003) studied the Bayesian Monte Carlo and Markov Chain Monte Carlo methods applied to probabilistic ecological modelling, concluding that Bayesian Monte Carlo is very inefficient while the Markov Chain Monte Carlo has a fast convergence.

In the finance domain, Boyle et al. (1997) reviewed the Monte Carlo method's application to security pricing problems, focusing on efficiency improvements. Juneja (2010) explained Monte Carlo methods for pricing American options and portfolio risk measurement. Finally, Whiteside (2008) applied Brownian-walk Monte Carlo simulation in forecasting, which forecasts commodities price changes.

In the industrial domain, it is very often required to estimate the uncertainty related to pH measurements for processes such as food, cosmetics, textiles, chemicals, etc. Cristancho and Castillo (2018) discussed the application of Monte Carlo methods to meet that requirement. Kim and Hur (2020) applied the Monte Carlo method to account for the uncertainty associated with variable generation resources (VGR), such as wind and solar renewable energy resources, which can impact the increasing use of these energy sources into systems such as electric vehicles.

## 4.8 Summary

Chapter four starts with a brief introduction to the origins and rationale of the Monte Carlo simulation. A few simple applications of Monte Carlo are explained with codes developed using the R language. An application of Monte Carlo for project management is discussed in detail. In Monte Carlo simulation, one vital aspect to understand is when to stop the simulation or how to define the number of simulations required to achieve a given level of accuracy. For that, we dedicate a section to discussing this aspect. The Monte Carlo workflow is explained in detail, including limitations and challenges. One of the crucial aspects of the Monte Carlo workflow is the definition of the probability distribution functions required in each problem to capture the uncertainty in the parameters we intend to model. Several methods are described, and emphasis is given to the maximum likelihood estimators method, which allows the adjustment of known distribution probability parameters to actual data; a few examples are shown of the maximum likelihood estimators method. Random number generation is discussed with a focus on the inverse transformation method, and a few examples are included. A brief discussion on the acceptance-rejection method for generating a random number is introduced. A complete example of Monte Carlo simulation is applied to an assessment of oil in place in an oil reservoir. An important random sampling technique called Latin Hypercube sampling is introduced, and its rationale and need are explained. The chapter ends with the advantages and disadvantages of Monte Carlo simulation, and applications in several domains are referred to.

# References

Boyle, P. H., Broadie, M., & Glasserman, P. (1997). Monte Carlo methods for security pricing. *Journal of Economic Dynamic and Control, 21*(1997), 1267–1321. Published by Elsevier.

Chernoff, H., & Lehmann, E. (1954). The use of Maximum likelihood estimates in $\chi 2$ test for goodness of fit. *Annals of Mathematical Statistics, 25*, 579–586.

Cristancho, R., & Castillo, A. (2018). Uncertainty estimation of a primary pH-measurement system by the Monte Carlo simulation method. *Journal of Physics*: *Conference Series, 1119*(012012). https://doi.org/10.1088/1742-6596/1119/1/012012

Eckhardt, R. (1987). Stan Ulam, John von Neuman, and the Monte Carlo Method. *Los Alamos Science, Special Issue*, 131–143.

Evans, G., & Jones, B. (2009). The application of Monte Carlo simulation in finance, economics and operations management. In *Proceedings of the 2009 WRI World Congress on Computer Science and Information Engineering*, Los Angeles, CA, USA, 2009 (pp. 379–383). https://doi.org/10.1109/CSIE.2009.703

Firestone, M., Fenner-Crisp, P., Barry, T., Bennett, D., Chang, S., Callahan, M., Burke, A., Michaud, J., Olsen, M., Cirone, P., Barnes, D., Wood, W., & Knott, S. (1997). Guiding principles for Monte Carlo. Retrieved March 8, 2018, from U.S. Environmental. Protection Agency (EPA): Risk Assessment Review: https://www.epa.gov/sites/production/files/2014-11/documents/mon tecar.pdf

Furness, P. (2011). Applications of Monte Carlo simulation in marketing analytics. *Journal of Direct, Data and Digital Marketing Practice, 13*, 132–147. https://doi.org/10.1057/dddmp.2011.25

Hammersley, J., & Handscomb, D. (1975). *Monte Carlo methods*. Methuen's Monographs on Applied Probability and Statistics. General editor: M. S. Bartlett.

Juneja, S. (2010). Monte Carlo methods in finance: An introductory tutorial. In B. Johansson, S. Jain, J. Montoya-Torres, J. Hugan, & E. Yucesan (Eds.), *Proceedings of the 2010 Winter Simulation Conference* (pp. 95–103).

Kim, S., & Hur, J. (2020). (2020) A probabilistic modeling based on Monte Carlo simulation of wind powered EV charging stations for steady-state security analysis. *Energies, 13*(5260), 1–13. https://doi.org/10.3390/en13205260

Kok, M., Kaya, E., & Akin, S. (2006). Monte Carlo simulation of oil fields. *Energy Sources*, Part B, *1*, 207–211. Taylor & Francis Group, LLC. https://doi.org/10.1080/15567240500400770

Kosova, R., Shehu, V., Naco, A., Xhafaj, E., Stana, A., & Ymeri, A. (2015). Monte Carlo simulation for estimating geologic oil reserves. A case study from Kucova Oilfield in Albania. *Muzeul Olteniei Craiova. Oltenia. Studii şi comunicări. Ştiinţele Naturii, 31*(2), 20–25.

Metropolis, N. (1987). The beginning of the Monte Carlo method. *Los Alamos Science, Special Issue, 1987*, 125–130.

Metropolis, N., & Ulam, S. (1949, September). The Monte Carlo method. *Journal of the American Statistical Association, 44*(247), 335–341.

Platon, V., & Constantinescu, A. (2014). Monte Carlo method in risk analysis for an investment project. *Procedia Economics and Finance, 15*(2014), 393–400. Published by Elsevier.

Song, Q., Craig, S., & Borsuk, M. (2003). On Monte Carlo methods for Bayesian inference. *Ecological Modelling, 159*, 269–277. Published by Elsevier.

Taha, H. (2006). *Operations research: An introduction* (8th ed.). Pearson Prentice.

Tenekedjiev, K., Nikolova, N., & Kolev, K. (2011). Applications of Monte Carlo simulation in modelling of biochemical processes. In C. J. Mode (Ed.), *Applications of Monte Carlo methods in biology, medicine and other fields of science* (pp. 57–76). ISBN: "-953-307-427-6. InTech. http://www.intechopen.com/books/applications-of-monte-carlo-methods-in-biology-medicine-and-other-fieldsof-science/applications-of-monte-carlo-simulation-in-modelling-of-biochemical-processes

Whiteside, J. (2008). A practical application of Monte Carlo simulation in forecasting. In *Proceeding of the 2008 AACE International Transactions* (pp. 1–12).

Wright, D. (2019). *Basics of Monte Carlos simulation.* Geant4 Tutorial at Sao Paulo, National Accelerator Laboratory. https://indico.cern.ch/event/776050/contributions/3237662/attachments/178 1016/2897449/MCIntroSaoPaulo.pdf

# Chapter 5
# Assessing the Importance of the Uncertainties: Design of Experiments

**Objective**

In this chapter, we describe the experimental design method. This methodology systematically and efficiently assesses the system outcomes (reservoir production) at given values of the input variables (permeability, contacts, oil rate, etc.). It is a valuable technique to estimate the uncertain variables with the highest impact on the project value. It is used, in this book, to steer the data acquisition actions that provide the most significant increase in the project value: the value of information.

## 5.1 Introduction to Experimental Design

When we are faced with a decision problem under uncertain conditions, there are situations in which data can be acquired to potentially reduce the uncertainty of the project and increase its value. However, identifying one data acquisition action that supports the decision process by increasing today's value of a future project is not enough. A project should be considered a complete entity. Therefore, data acquisition aims to identify several data gathering actions that provide the most positive impact on the project's value. This statement means that we are not looking for a data acquisition action that can impact a variable's uncertainty; we are, instead, looking for the data acquisition action that has the more significant positive impact on the project's value.

In order to do this, the Design of Experiment (DOE) technique is beneficial; it looks for all the uncertain variables and ranks them from the one that impacts the most on a project's value to those with minor impact. Secondly, we search for the complete set of data acquisition actions associated with the project. From that set, the action which has the highest impact on the uncertainty is performed first and so on until the project is finally accepted or rejected.

© The Author(s), under exclusive license to Springer Nature Switzerland AG 2022
M. J. Vilela and G. F. Oluyemi, *Value of Information and Flexibility*,
Petroleum Engineering, https://doi.org/10.1007/978-3-030-86989-2_5

DOE consists of a methodology for applying statistics to experimentation by making changes in the input variables of a system or process and measuring the corresponding changes in the output variables.

Several designs are available for doing experiments within DOE (factorial, fractional factorial, central composite, Box-Behnken, etc.). However, for the scope of this book, and with the primary objective being decision problems (especially data acquisition problems), we concentrate on the factorial design. The reason for this focus is that, for data acquisition problems, the use of DOE is associated with screening type analysis for identifying the main factors affecting the response. Therefore, those kinds of analyses are made using factorial designs.

DOE has been extensively used for reservoir uncertainty analysis (White & Royer, 2003) and response surface applications in reservoir simulations studies (Amudo et al., 2008; Moeinikia & Alizadeh, 2012).

For a further review on the topic of DOE, we suggest Montgomery (2013).

## 5.2  Background of Design of Experiments

An experiment is a process whose outcome is not known with certainty. This definition means that an experiment can produce different results. The set of all possible outcomes of an experiment is called the *sample space*, denoted by $S$, and each independent outcome is a *sample point*.

DOE is a structured manner of conducting and analysing tests for evaluating and ranking the factors that affect the response variable. DOE uses statistical tools for system performance optimisation by defining the importance of specific processing and variables and their interactions and establishing how to combine them to get optimum outcomes. Also, DOE is used to develop new processes or products for optimising existing ones when more than one variable is involved.

During DOE, the analyst defines the setting levels or values of the input variables and the combination of factors in which the experiments should be run.

DOE was developed during the 1920s and 1930s by *Sir Ronald Fisher* at the agricultural research station of Rothamsted Agricultural Field Research Station in North London, England (Fisher, 1935), as discussed by Telford (2007). Those experiments were conducted to investigate the effect of several fertilisers on different plots amongst the other factors that impact crop growth, such as soil conditions, moisture content of the soil, temperature, etc. DOE allowed differentiation of the effect of fertiliser on the impact of the other factors.

In his work, Fisher (1935) showed that a valid conclusion could be obtained efficiently from experiments with natural fluctuations in *known nuisance variables* such as temperature, rainfall, humidity, and soil conditions (Montgomery, 2013). These are known nuisance variables and cause systematic biases; DOE also considers *unknown nuisance variables* that cause random variability in the results or noise.

Even though the first application of DOE was in the agricultural domain, it has proved helpful in many other fields. After Fisher's work, the next milestone in experimental design came from Box and Hunter's work (1957), who applied DOE to industrial experiments and developed the Response Surface Methods (RSM). RSM is a particular subject that will not be discussed because it is beyond the objective of this book. However, due to its significance and interest, it is recommended to the curious reader (Oehlert, 2010).

In industrial research, development, and production DOE has been used for:

- Optimisation of analytical instruments.
- Screening and identification of critical factors.
- Optimisation of manufacturing processes.
- Robustness testing of products.
- Reducing process design and development time.
- Improving profits and return on investment.
- Formulation experiments.

Today, DOE is used in manufacturing processes and engineering domains to explore how the input variables to processes or systems impact the output performance variables. Examples would be defining the quantity and type of fertiliser that provides the highest rate of tomato production or determining the optimal temperature for operating a reactor to produce the optimum quality of a particular chemical compound. In addition, DOE has been used in many industries, such as those producing automobiles, semiconductors, medical devices, and chemical products and processes like surgery and advertising strategies (Durakovic, 2017).

## 5.3  Design of Experiments Methodology

During an experiment, a series of runs are carried out. Changes are made to the input variables of a system or process so conclusions can be reached on how those changes are related to the observed variability in the output response.

When an equation relates the variables in a system, we call that a *mechanical model*. In contrast, when the variables in a system are described by experimentation, we call an *empirical model*.

Experiments are carried out to understand the performance of a process or system; Fig. 5.1 shows a schematic of a generic system or process.

At a high level, the system or process is described by a set of input variables (materials, properties, etc.), factors that can be controllable (pressure, etc.) or uncontrollable (weather conditions, etc.), and a set of outputs or response variables (production, revenues, etc.). The system is the entity that combines the input variables and the factors to produce the outcomes.

DOE is used for:

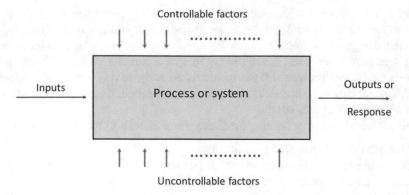

**Fig. 5.1** The general schematic of a system or process

- *Comparison*: comparing multiple designs, based on the input variable options, to choose the best one. For example, when selecting different providers of circuits for computers, choose the one that achieves fast data transmission.
- *Screening*: (i) explore many factors to reveal whether they influence the responses and (ii) identify their appropriate ranges. Screening allows selecting the variables that most affect the system's performance from a large number of variables. Screening frequently considers only two levels of each factor; it is sometimes called sensitivity analysis.
- *Transfer function*: Typically, a screening design is applied first to identify the relevant factors. Once the most important variables have been identified, their impact on system performance can be analysed using the transfer function, the relation between the critical input variables, and the output response. The designs used for transfer function identification have several levels on each factor called response surface methods.
- *Optimisation*: (i) to predict the response's values for all possible combinations of factors within the mathematical region and (ii) to identify an optimal experimental point. Usually, it is difficult to find a point that fulfils all responses, which means a trade-off between the conflicting goals is required. By optimisation, we can improve the efficiency, quality, and reliability of the system.
- *Robustness*: (i) to ascertain that the method is robust to small fluctuations in the levels of the factors and (ii) to understand how to alter the bounds of the factors so that robustness may still be claimed in cases where non-robustness is detected. The system is robust when it is not sensitive to noise, such as environmental factors and uncontrolled factors.

Typically, several input variables are involved in a process, and one or more output performance variables need to be estimated, and these are the responses of interest. Each input variable can have one or more 'levels' or values significant to the given process under study. Each set of values for the input variables are a *treatment* that will generate a response. Even when using the same input variable, the responses are not always the same due to known or unknown nuisance variables.

In an experiment, the analyst intentionally changes input variables to measure the corresponding changes in the response of interest. When several input variables impact a particular characteristic of a process (the response of interest), several treatments can be carried out to characterise the process in terms of the most important input variables and the ranges of their values that generate the optimum result for the response of interest.

In an experiment, not all the variables have the same impact on the system's outcomes; indeed, some have a more significant impact than other variables. Therefore, one of the experiments' objectives is to determine which variables have a more substantial influence on the system's performance, so it can be optimised to work in the most beneficial conditions.

The DOE's objective is to design the experiments to draw reliable and robust conclusions efficiently, effectively, and economically. Therefore, DOE defines which changes should be made in the input variables to obtain valuable information for understanding and maximising the process under consideration. Different input variables have a different impact on the system's outcome. Using statistical methods, DOE determines which variables have the most significant impact and which variables are less relevant. Some important definitions related to DOE are provided:

*Treatments*: refers to the processes we want to compare. Examples are different temperatures in a reactor in chemical engineering, amount of fertiliser in agronomy, amount of one compound in a medical drug, etc.

*Experimental units*: the things to which the treatments are applied. Examples of experimental units are a group of reactors in a factory, a plot for growing potatoes, a group of people testing a medical drug.

*Effect*: it is the difference in the average response between two different levels of a given factor or between two experimental runs. A positive effect increases as the independent factor changes from its low to high value; an adverse effect decreases as the independent factor increases from its minimum to maximum.

*Factor effect*: this is the average response for an independent factor at its high value, compared to the average response for the same factor at its low value.

*Interactions* occur when one factor's effect on the response is different at different levels of the other factors.

*Analysis of variance (ANOVA)*: this is a mathematical procedure for comparing the response data with the error data and determining whether an independent variable or their interactions are significant. ANOVA uses the sum of squares for comparing effects with the error variance to determine statistical significance.

*Experimental space*: this is the region defined by the high and low levels of an experiment's independent factors.

*Responses*: outcomes result from applying a treatment to an experimental unit measured in the experiment; an experiment can have more than one response. Examples of experimental responses are the time to produce a given amount of a chemical product, the number of potatoes per square metre, and the relief of a medical problem after taking the test drug.

*Factor level*: input variables can be qualitative or quantitative. For quantitative factors, we should decide the range of values, the levels (discrete values relevant to the study process), and how they will be measured and controlled during the experiment; examples are temperature and pressure. Qualitative factors are discrete and not numeric, such as raw materials, type of supplier etc. Level refers to a specific value of the factor that will be examined in the experiment.

*Experimental error*: random variation is present in an experimental result; when the same treatment is applied to different experimental units, the products are different and, very often, when the same treatment is carried out several times, the same experimental unit gives different results.

*Measurement units*: these are the actual objects on which the response is measured; an example would be, while the experimental unit is a field plot, the measurement unit is the area dedicated to growing potatoes.

*Factors*: these are the variables or elements of the process or system that change in the treatments. Examples are the speed of a process, the temperature of a device, the concentration of fertiliser, etc.

*Confounding*: this refers to the combined impact of two or more factors in one measured effect; this occurs when the effects of one factor or treatment cannot be separated from another factor or treatment. Effects that are confounding are called aliases.

*Controlled experiment*: an experiment is controlled when the analyst assigns treatments that they decide.

*Control treatment*: this is the base treatment that acts as the comparison case to evaluate proposed treatments.

*Replication*: this is the number of times an experiment is repeated at identical factor levels. Replication allows the determination of the error that occurs in the experiment.

In Sects. 5.3 and 5.4, we describe two unique designs, the One-Variables-at-a-Time, and the Randomised complete block designs. These designs do not use the statistical techniques used in the others. More sophisticated designs are discussed from Sect. 5.5 onward.

## 5.4  One-Variable-At-a-Time Design

Many analysts use the most straightforward experimental design, One-Variable-At-a-Time (OVAT), sometimes called *One-Factor-At-a-Time* (OFAT). In the OVAT method, a base value is set for all variables, and then, one variable is varied over its

**Table 5.1** Example 5.1: OVAT

| Treatment | A | B | Average response |
|-----------|-----|-----|------------------|
| 1 | A1 | B1 | 70 |
| 2 | A1 | B2 | 55 |
| 3 | A1 | B1 | 70 |
| 4 | A2 | B1 | 50 |

range, while the other variables are kept fixed at their base value. Then, all combinations of variables are carried out, and the set of outcomes are generated and analysed. Typically, several types of plots are used to evaluate the effect of the different variables on the response, and the optimal selection of variables can be made. OVAT strongly depends on guesswork, luck, experience, and intuition for its success.

The main disadvantages of OVAT are (i) it requires many experiments, which means ample resources and time to obtain a limited amount of information, and (ii) it fails to consider the possible interaction between variables. The interaction between variables occurs when the effect on the outcomes from one variable depends on another variable's values. As a result, OVAT designs are unreliable, inefficient, and time-consuming. Essentially, they can produce wrong conclusions, as explained in the following example.

**Example 5.1: OVAT** Assume we have a process depending on two input variables, A and B. The output performance variable is C. Consider that A and B have two levels each, A1 (base value) and A2, and B1 (base value) and B2. For each treatment, two trials (replications) were made. Table 5.1 summarises the design and the average responses for the corresponding trials.

Based on these results, the higher response corresponds to the treatment when A is at level A1 and B is at level B1. From the data shown in Table 5.1, the difference between treatments 1 and 2 measures the effect of variable B, which is 21% $((70 - 55) \cdot 100\%/70)$, while the difference between treatments 3 and 4 gives the effect of variable A, which is 29% $((70 - 50) \cdot 100\%/70)$.

However, from this data, we do not know the effect of variable B at level A2 or A's impact at level B2, which can differ from the one shown in Table 5.1. Besides, we do not know the possible effects of the interactions between A and B. These facts mean that OVAT can result in misleading conclusions.

Conducting experimentation in this manner is inefficient because it requires significant resources to obtain a limited amount of information about the process. Nevertheless, Tanco (2008) found that 80% of the companies that carry out a study in the Basque Country, Spain, use OVAT as an experimentation strategy.

A better approach for experimentation is the *factorial experiment design,* in which variables are varied together, as discussed in Sect. 5.6.

## 5.5   Randomised Complete Block Design

The randomised complete block design (RCBD) experimentation method applies
when one factor is considered the main one responsible for the outcome. The experiment's focus is to get the influence of this factor in the outcome of the system; this
factor is called the *primary factor*. The remaining factors are called nuisance factors.
RCBD uses the blocking technique on all factors except the primary factor. It makes
a set of runs with all the values or levels of the primary factor, keeping the nuisance
factors constant; this process is repeated for all possible sets of nuisance values. As
part of the randomisation, the runs are made in random order.

**Example 5.2: RCBD**   Assume that we have two factors in an experiment (A and B),
where B is the primary factor, and A is the nuisance factor. A has three levels: $A_1$,
$A_2$ and $A_3$ and B also has three levels: $B_1$, $B_2$ and $B_3$. Considering two replications,
Table 5.2 summarises the design of this experiment.

Table 5.2 shows the RCBD design, where green highlights the set of runs and
light-red shows the replication.

**Table 5.2**  Example 5.2: RCBD

| Blocks | Runs | Nuisance factor | Primary factor |
|---|---|---|---|
| Block 1 | 1 | $A_1$ | $B_1$ |
| | 2 | $A_1$ | $B_2$ |
| | 3 | $A_1$ | $B_3$ |
| | 4 | $A_1$ | $B_1$ |
| | 5 | $A_1$ | $B_2$ |
| | 6 | $A_1$ | $B_3$ |
| Block 2 | 7 | $A_2$ | $B_1$ |
| | 8 | $A_2$ | $B_2$ |
| | 9 | $A_2$ | $B_3$ |
| | 10 | $A_2$ | $B_1$ |
| | 11 | $A_2$ | $B_2$ |
| | 12 | $A_2$ | $B_3$ |
| Block 3 | 13 | $A_3$ | $B_1$ |
| | 14 | $A_3$ | $B_2$ |
| | 15 | $A_3$ | $B_3$ |
| | 16 | $A_3$ | $B_1$ |
| | 17 | $A_3$ | $B_2$ |
| | 18 | $A_3$ | $B_3$ |

## 5.6  Design of Experiments Workflow

DOE is the procedure for assigning treatments to the experimental units to verify whether the difference between treatments is true or due to random variability and establish a tendency between variables.

The fundamental principles of DOE are processes applied to the experiment design to solve the problems occurring for the presence of known and unknown nuisance factors; these principles are:

(i)   Randomisation
(ii)  Replication
(iii) Blocking
(iv)  Orthogonality.

*Randomisation*: refers to the random order in which the experiments are performed. Randomisation ensures that the run's conditions neither depend on the previous run nor anticipate the next run's conditions. When randomisation is implemented in DOE, bias is avoided. To avoid bias, due to external known or unknown variables (e.g. temperature, altitude, etc.), during the comparison of two different processes (A and B), the experiment should be run randomly, selecting the order in which process A and B are evaluated. Randomisation is the use of probabilistic mechanics for the assignment of treatment to units. Randomisation is used to incorporate analysis factors such as operator errors, fluctuations in ambient conditions, and raw material variations, which could adversely affect the experiment's results. Randomisation removes all sources of difficulties to control variations in the experiments: it averages the noise effects by assuring that all factor levels have the same chance of receiving the impact's noise factors.

*Replication*: Replication is a complete or partial repetition of the experiment to obtain a more precise result and estimate the experimental error. It increases the experiment's precision by increasing the signal-to-noise ratio when noise is due to uncontrollable nuisance variables. Replications allow us to obtain (i) a more accurate estimate of the experimental error, (ii) a more precise estimate of the interaction between factors and, (iii) a lower experimental error and higher precision. Of course, replication can significantly increase experimental time and cost; a trade-off between accuracy and cost is considered.

*Blocking*: is the procedure by which a known systematic bias effect is isolated, preventing it from hiding the main impact. During blocking, the experiments are arranged in similar groups. This arrangement reduces the sources of variability, and precision is improved. Similar experimental trials are set in blocks characterised by a common variable, raw material, operator, or vendor when blocking is applied. By removing the effect of known nuisance factors, this method increases the experiment's precision; when blocking is implemented, blocks characterised by a known nuisance are built; randomisation is then applied to the blocks. An example of a blocking application occurs when experiments are conducted using several raw materials, and batch-to-batch variability is expected. The blocking method limits complete randomisation, but randomisation is applied inside each block. Observations gathered in the same block have the same operational conditions. By definition, blocking increases the precision of the experiment.

*Orthogonality*: this means that the factor effects are uncorrelated, and so, they vary independently from each other. In an orthogonal design, the result of the experiment can be summarised using average values.

The workflow for performing DOE consists of seven steps:

(i)     Defining the problem and determining the experiment's objectives: it should be well-defined and understood by all people involved. The scope is clearly explained, and the relevance of the outcome is understood. The relation between the problem and the experiment's objectives should be established.

(ii)    Choose responses: responses are the outcome of the experiments. A given experiment can have one or more responses.

(iii)   Choose factors and levels: factors are the input in the experiments responsible for the responses obtained in the experiments; factors are studied by experimentation. The levels or settings of the factors are the values of the factors used in the experiments, and they are constrained to the range of values suitable for them.

(iv)    Choose experimental design: the objective of the experiment indicates the design that is used. As discussed below, experiments can be undertaken for comparison, screening, transfer function, or optimisation, and each one defines the type of design. The number of factors and the corresponding levels also intervene in the selection of the design.

(v)     Perform the experiment: different values of the factors should be used for each experiment following a specified sequence. Typically, a design matrix is used to guide the experiments that are to be conducted.

(vi)    Analyse data: Analysis of Variance (ANOVA) and regression analysis is the standard statistical procedure for analysing the experiments. Statistical methods, engineering knowledge, and common sense work together to provide conclusions on the essential factors.

(vii)   Draw conclusions and recommendations: based on the analysis, conclusions should be made. Usually, graphical methods are used for presenting the results.

The application of these steps secures a satisfactory implementation of experiments to obtain the desired results.

## 5.7  Full Factorial Design

*The full factorial design* consists of all possible combinations of levels for all factors. The full factorial design method estimates each factor's effects and all the interaction effects between factors. The full factorial design does not distinguish between primary and nuisance factors a priori. The simplest one is the two-level factorial design where, for $k$ factors and $L = 2$ levels, the sample size is $N = 2^k$. Typically, in the early stages of experimental work on a particular problem, it is widespread to use full factorial design with two levels, especially if the number of parameters is less than 5. This design assumes that the response is approximately linear over the range of the factors.

For $k$ factors, each with $n$ levels, the total number of experiments is $N = n^k$. All possible combinations of the factors' levels are investigated in a complete trial or replication in a complete factorial design.

The two levels are usually called 'h' and 'l' (high and low) or '+1' and '−1'. Because all combinations are explored, the effect of each factor over the response is not confounded with other factors. In some circumstances, the central point is added in the sample space: the run in which all the factors are set at their average value between the low and high values. This central value is usually called 'm' or '0'.

Factorial designs have, by definition, internal replication because the effect of each factor is the difference between the averages of half of the runs, those in which the factors are evaluated at the 'high' levels, and the other half of the runs evaluated at 'low' levels. These designs where the distribution of runs is half-and-half, between 'high' and 'low' values of the factors, are called balanced designs.

The factorial design method performs experiments to estimate each factor's effect and the effect of the interaction between factors. The factorial design can be represented by a hyper-cube of $n$-dimensions, where $n$ is the number of factors. This design runs the experiment at each vertex of the cube, where each vertex corresponds to different combinations of factor levels. The number of experiments in a full factorial design is given by $n^p$ where $n$ is the number of factors and $p$ is the number of levels per factor.

The $2^k$ designs are often called screening designs. These designs are built to explore many factors, using the minimal number of levels, two. By screening, we mean the process of screening a large number of factors that may be important in the system under investigation, intending to choose those that are more important to the system's response.

Figure 5.2 shows a 3-dimensional representation of the factorial design with 3-factors and 2-levels $(2^3)$ where the two levels of each factor are labelled $+ 1$ and $- 1$ and, each vertex corresponds to one experiment.

Because, in the full factorial design, the number of combinations (and experiments) increases very quickly with the number of factors and (or) the number of levels. In practical applications where the number of factors or levels is large, we should consider different designs.

**Example 5.3: Simple Full Factorial Design, 2-Factors, 2-Levels** The simplest $2^k$ design is when $k = 2$. The factors are usually labelled with capital letters ($A$, $B$, $C$, etc.). These may be the concentration of a chemical product, temperature, humidity, or concentration of a reactant, etc.

**Fig. 5.2** 3 Factorial design
with 2-levels per factor

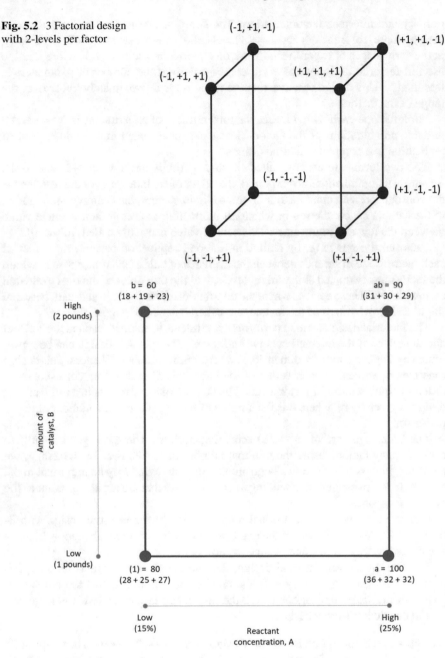

**Fig. 5.3** Example 5.3: screening design 2-factors, 2-levels

Figure 5.3 shows an example of a $2^k$ design, described by Montgomery (2013), which uses Yates' notation. In this case, we have four treatments with three replications, consequently, three observations per treatment.

Following Yates' notation, '(1)' denotes that both factors are at the low level, 'a' indicates that $A$ is at its high and $B$ is at its low level, 'b' signifies that $B$ is at its high and $A$ is at its low, and 'ab' stand for both $A$ and $B$ are at their high level.

The main effect of $A$ is the average of the effect of $A$ at the high level of $B$ $\left(\frac{[ab-b]}{n}\right)$ and the effect of $A$ at the low level of $\left(\frac{[a-(1)]}{n}\right)$, as shown in Eq. 5.1,

$$A = \frac{1}{2}\left(\frac{[ab-b]}{3} + \frac{[a-(1)]}{3}\right) = 8.33 \tag{5.1}$$

In the same way, the main effect of $B$ is the average of the effect of $B$ at the high level of $A$ and the effect of $B$ at the low level of $A$, Eq. 5.2,

$$B = \frac{1}{2}\left(\frac{[ab-a]}{3} + \frac{[b-(1)]}{3}\right) = -5.00 \tag{5.2}$$

The interaction effect $AB$ is the average difference between the effect of A at the high level of $B$ and the effect of $A$ at the low level of $B$, Eq. 5.3,

$$AB = \frac{1}{2}\left(\frac{[ab-b]}{3} - \frac{[a-(1)]}{3}\right) = 1.67 \tag{5.3}$$

The effect of $A$ (the reactant concentration) is positive because increasing $A$ increases the yield. The effect of $B$ (the catalyst) is negative because increasing $B$ decreases the yield. The interaction effect, although positive, is small compared with the main effects.

**Example 5.4: Full Factorial Design with 4-Factors, 2-Levels, Single Replication**
In this example, we consider that the amount yield of a product depends on four factors A, B, C, and D. Each factor has two levels, high (+) and low (−). Table 5.3 summarises the design and the yield resulting from each treatment.

The runs are made in random order. The objective of this experiment is to obtain the combination of factors that produces a higher yield.

The Pareto plot of the 16 experiments (Fig. 5.4) indicates that, at a 5% significance level, the main factors contributing to the response are B, A, and C in that order; also, no interaction factors are relevant.

The same conclusion reached by the Pareto plot can also be shown using the Normal plot, where the effects that lie along the line are negligible, whereas the significant effects are far from the line, as shown in Fig. 5.5.

Figures 5.6 and 5.7 show the main and the interaction effect plots, respectively; these figures confirm that the main effects of factors A and B are the most important ones, and the interaction effects are not significant (Table 5.4).

These effects are used to define the regression equation of this example, Eq. 5.4,

**Table 5.3** Matrix design for Example 5.4

| Run number | Factor | | | | Run label | Filtration rate |
|---|---|---|---|---|---|---|
| | A | B | C | D | | |
| 1 | −1 | 1 | −1 | −1 | 3 | 74 |
| 2 | −1 | 1 | 1 | 1 | 15 | 84 |
| 3 | 1 | −1 | 1 | −1 | 6 | 87 |
| 4 | 1 | −1 | −1 | −1 | 2 | 81 |
| 5 | 1 | 1 | −1 | −1 | 4 | 90 |
| 6 | −1 | 1 | 1 | −1 | 7 | 86 |
| 7 | 1 | −1 | −1 | 1 | 10 | 69 |
| 8 | 1 | −1 | 1 | 1 | 14 | 88 |
| 9 | −1 | −1 | −1 | 1 | 9 | 45 |
| 10 | 1 | 1 | 1 | 1 | 16 | 120 |
| 11 | −1 | 1 | −1 | 1 | 11 | 92 |
| 12 | −1 | −1 | 1 | 1 | 13 | 61 |
| 13 | 1 | 1 | 1 | −1 | 8 | 105 |
| 14 | −1 | −1 | 1 | −1 | 5 | 65 |
| 15 | −1 | −1 | −1 | −1 | 1 | 44 |
| 16 | 1 | 1 | −1 | 1 | 12 | 99 |

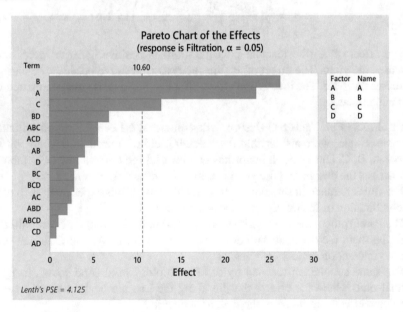

**Fig. 5.4** Example 5.4: Pareto plot for the experiment

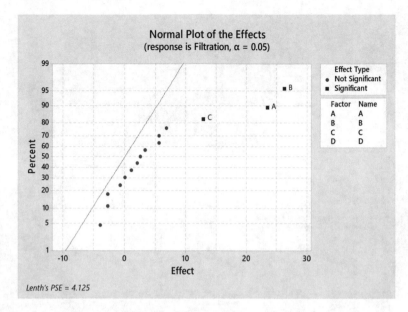

**Fig. 5.5**  Example 5.4: normal plot for the experiment

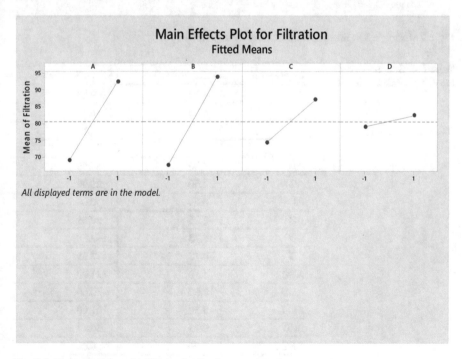

**Fig. 5.6**  Example 5.4: main effects plot for the experiment

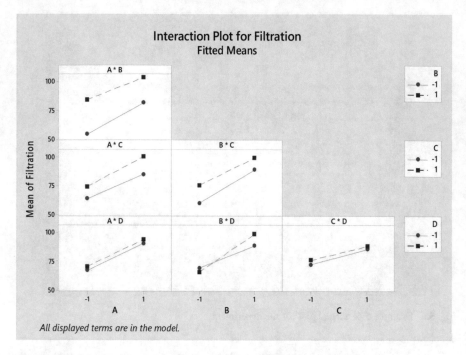

**Fig. 5.7**  Example 5.4: interaction effects plot for experiment

**Table 5.4**  Summary of all the effects of Example 5.4

| Model term | Effect estimate | Sum of squares |
|------------|-----------------|----------------|
| A          | 23.50           | 2209.00        |
| B          | 26.25           | 2756.25        |
| C          | 12.75           | 650.25         |
| D          | 3.25            | 42.25          |
| AB         | −4.00           | 64.00          |
| AC         | 2.50            | 25.00          |
| AD         | 0.00            | 0.00           |
| BC         | −2.75           | 30.25          |
| BD         | 6.75            | 182.25         |
| CD         | −0.75           | 2.25           |
| ABC        | 5.50            | 121.00         |
| ABD        | 2.00            | 16.00          |
| ACD        | 5.50            | 121.00         |
| BCD        | −2.75           | 30.25          |
| ABCD       | 1.00            | 4.00           |

**Table 5.5** Full factorial design with 3-factors and 2-levels each

| Experiment number | Factor levels | | | Response variables | Factor interactions | | | |
|---|---|---|---|---|---|---|---|---|
| | $X_1$ | $X_1$ | $X_1$ | | $X_1X_2$ | $X_1X_3$ | $X_2X_3$ | $X_1X_2X_3$ |
| 1 | $-1$ | $-1$ | $-1$ | $y_{l,l,l}$ | $+1$ | $+1$ | $+1$ | $-1$ |
| 2 | $-1$ | $-1$ | $+1$ | $y_{l,l,h}$ | $+1$ | $-1$ | $-1$ | $+1$ |
| 3 | $-1$ | $+1$ | $-1$ | $y_{l,h,l}$ | $-1$ | $+1$ | $-1$ | $+1$ |
| 4 | $-1$ | $+1$ | $+1$ | $y_{l,h,h}$ | $-1$ | $-1$ | $+1$ | $-1$ |
| 5 | $+1$ | $-1$ | $-1$ | $y_{h,l,l}$ | $-1$ | $-1$ | $+1$ | $+1$ |
| 6 | $+1$ | $-1$ | $+1$ | $y_{h,l,h}$ | $-1$ | $+1$ | $-1$ | $-1$ |
| 7 | $+1$ | $+1$ | $-1$ | $y_{h,h,l}$ | $+1$ | $-1$ | $-1$ | $-1$ |
| 8 | $+1$ | $+1$ | $+1$ | $y_{h,h,h}$ | $+1$ | $+1$ | $+1$ | $+1$ |

$$Yield = 80.630 + 11.750 * A + 13.130 * B + 6.375 * C + 1.625 * D$$
$$- 2.000 * AB + 1.250 * AC + 0.000 * AD - 1.375 * BC$$
$$+ 3.375 * BD - 0.375 * CD + 2.750 * ABC + 1.000 * ABD$$
$$+ 2.750 * ACD - 1.375 * BCD + 0.500 * ABCD \qquad (5.4)$$

**Example 5.5: Full Factorial Design, 2-Factors, 3-Levels** Table 5.5 shows a full factorial design with three factors and two levels per factor. This experiment consists of $2^3 = 8$ runs; it also includes the interaction of the factors.

The main interaction $M$ of a variable $X$ is the difference between the average response at the high-level samples and the average response at the low-level samples. For the data in Table 5.5, the main interaction for the factor $X_1$ is given by Eq. (5.5)

$$M_{X_1} = \frac{y_{h,l,l} + y_{h,l,h} + y_{h,h,l} + y_{h,h,h}}{4} + \frac{y_{l,l,l} + y_{l,l,h} + y_{l,h,l} + y_{l,h,h}}{4} \qquad (5.5)$$

For the other two factors, similar equations apply. The interaction effect of two or more factors is the difference between the average responses at the high and low levels in the interaction column. For example, the interaction effect between $X_1$ and $X_2$ is computed as per Eq. (5.6),

$$M_{X_1,X_2} = \frac{y_{l,l,l} + y_{l,l,h} + y_{h,h,l} + y_{h,h,h}}{4} + \frac{y_{h,l,l} + y_{h,l,h} + y_{l,h,l} + y_{l,h,h}}{4} \qquad (5.6)$$

The main and the interaction effects provide an estimation of the impact of the factors in the response. The number of main and interaction effects in a $2^k$ full factorial design is $2^k - 1$, which is the design degree of freedom. The number of main effects in a factorial design $2^k$ is given by Eq. (5.7),

$$\binom{k}{1} = \frac{k!}{1!(k-1)!} = k \qquad (5.7)$$

The number of 2-interaction terms is given by Eq. (5.8):

$$\binom{k}{2} = \frac{k!}{2!(k-2)!} = \frac{k(k-1)}{2} \tag{5.8}$$

The number of $j$-interaction terms in a full factorial design $2^k$ is given by Eq. (5.9):

$$\binom{k}{j} = \frac{k!}{j!(k-j)!} \tag{5.9}$$

**Example 5.6: Experiment with a Single Factor with Replications** We assume that our experiment depends on one input parameter and results in one outcome. The input parameter can be set at four different levels. The experiment's objectives are to determine whether the parameter levels significantly impact the experiment's outcome and identify the levels that produce the optimum result. In other words, we wish to test for differences between the outcomes at all four levels of the parameter or confirm the equality of all four means. In this example, for each level, we conduct five trials or replications; the trials are carried out in random order.

Table 5.6 shows the results observed on the 20 experiments conducted and summarises the outcomes and average values.

ANOVA is the statistical method used to analyse the results of the experiment. In the Appendix of this chapter, we include a brief description of the ANOVA technique.

The statistic parameters of this experiment are shown in Table 5.7.

The value of $F$, with a level of significance of 0.05 and 3 degrees of freedom for the numerator and 16 degrees of freedom for the denominator, is 3.24. This value

**Table 5.6** Example 5.6: experiment with a single factor with replications

| Parameter level | Observed values | | | | | | Totals | Averages |
|---|---|---|---|---|---|---|---|---|
| | 1 | 2 | 3 | 4 | 5 | | $y_i$. | $\bar{y}_{i.}$ |
| A | 1,525 | 1,495 | 1,435 | 1,604 | 1,483 | | 7,542 | 1,508.40 |
| B | 1,489 | 1,501 | 1,604 | 1,521 | 1,517 | | 7,632 | 1,526.40 |
| C | 1,592 | 1,454 | 1,499 | 1,502 | 1,513 | | 7,560 | 1,512.00 |
| D | 1,543 | 1,601 | 1,492 | 1,624 | 1,511 | | 7,771 | 1,554.20 |
| | | | | | | $Y_{..}$ | 30,505 | |
| | | | | | | $\bar{Y}_{..}$ | 1,525.25 | |

**Table 5.7** Example 5.6: ANOVA evaluation

| Source of variability | Sum of squares | Degrees of freedom | Mean square | $F_0$ | P-value |
|---|---|---|---|---|---|
| Parameter | 6,494.55 | 3 | 2,164.85 | 0.74 | $P > 0.1$ |
| Error | 46,767.20 | 16 | 2,922.95 | | |
| Total | 53,261.75 | 19 | | | |

of $F$ is much larger than $F_0$ which indicates that there are no statistical differences between the treatments. In other words, the four levels of the parameter produce statistically similar results.

Suppose we reduce the level of significance by doubling it using 0.1 instead of 0.05. In that case, $F$ is equal to 2.46, meaning that even at that level of significance, there is no difference in treatment means. These results indicate that the $p$-value is greater than 0.1.

At present, several software packages can handle ANOVA calculations (SPSS, MINITAB, JASP, JAMOVI, etc.). We used the R language library AICcmodavg. Figure 5.8 shows the histogram of the data plotted using this library.

The output of the algorithm, when applied to this data set, is provided in Table 5.8.

The algorithm calculates the $p$-value as 0.543, confirming our preliminary assessment that $p$-value is greater than 0.1.

**Example 5.7: 2-Factor Experiments with 3-Levels, Including Replications** In this example, we consider an experiment with two factors, A and B, systematically varied in the experiments to produce a yield; each factor has three levels. The experiment's objective is to determine whether the several results observed due to the change in the levels of the factors are statistically significant. When the results are different, this could be due to any of the parameters or their interactions.

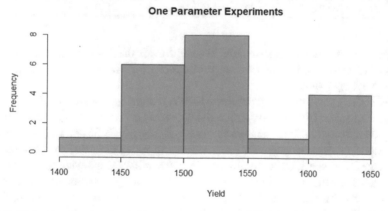

**One Parameter Experiments**

**Fig. 5.8** Histogram showing the yield values for the experiment described in Example 5.6

**Table 5.8**  Example 5.8: ANOVA evaluation using R language library AICcmodavg

|           | Df | Sum Sq. | Mean Sq. | F value | Pr(>F) |
|-----------|-----|---------|----------|---------|--------|
| Parameter | 3   | 6,495   | 2,165    | 0.741   | 0.543  |
| Residuals | 16  | 46,767  | 2,923    |         |        |

**Table 5.9** Example 5.7: 2-factor, 3-levels experimental results and average values

| Parameter A | Parameter B | | | | | | | | | | | | |
|---|---|---|---|---|---|---|---|---|---|---|---|---|---|
| | Low | | $y_{ij.}$ | $\bar{y}_{ij.}$ | Mid | | $y_{ij.}$ | $\bar{y}_{ij.}$ | High | | $y_{ij.}$ | $\bar{y}_{ij.}$ | $y_{i..}$ / $\bar{y}_{i..}$ |
| 1 | 12 | 10 | 48 | 12 | 7 | 10 | 40 | 10 | 6 | 12 | 41 | 10 | 129 / 11 |
| | 15 | 11 | | | 12 | 11 | | | 10 | 13 | | | |
| 2 | 16 | 12 | 60 | 15 | 14 | 16 | 52 | 13 | 11 | 8 | 55 | 14 | 167 / 14 |
| | 15 | 17 | | | 13 | 9 | | | 17 | 19 | | | |
| 3 | 14 | 11 | 50 | 13 | 17 | 15 | 57 | 14 | 10 | 12 | 39 | 10 | 146 / 12 |
| | 16 | 9 | | | 11 | 14 | | | 9 | 8 | | | |
| $y_{.j.}$ | | | 158 | | | | 149 | | | | 135 | | $y_{...}$ = 442 |
| $\bar{y}_{.j.}$ | | | | 13 | | | | 12 | | | | 11 | $\bar{y}_{...}$ = 12 |

**Table 5.10** Example 5.7: ANOVA evaluation

| Source of variation | Sum of squares | Degree of freedom | Mean squares | $F_0$ | P-Value |
|---|---|---|---|---|---|
| Parameter A | 60.39 | 2 | 30.19 | 3.51 | $0.05 < p < 0.025$ |
| Parameter B | 22.39 | 2 | 11.19 | 1.30 | $>0.05$ |
| Interaction | 36.44 | 4 | 9.11 | 1.06 | $>0.05$ |
| Error | 232.00 | 27 | 8.59 | | |
| Total | 351.22 | 35 | | | |

Table 5.9 shows the experimental results for all the combinations of the three levels of factors A and B and the four replications for each one; it also includes the sum and average values.

Table 5.10 shows the results of the ANOVA analysis, for Example 5.7.

The $F$ function for $\alpha = 0.05$ with 2 and 4 degrees of freedom in the numerator and 27 degrees of freedom in the denominators are $F_{0.05,2,27} = 3.35$ and $F_{0.05,4,27} = 2.73$.

Comparing these values with those estimated using the data, we conclude that parameter A is statistically significant (null hypothesis is rejected). In contrast, parameter B and the interaction factor are not significant (null hypothesis is accepted). So, based on this calculation with 5% precision, the measurements are different. However, if precision is reduced to 2.5%, $F_{0.025,2,27} = 4.24$ and $F_{0.025,4,27} = 3.31$, then no parameter is statistically significant.

The same calculation can be obtained using the R language library AICcmodavg. For this example, with two factors, Fig. 5.9 shows the histogram of the distribution of values.

The outcome of the code is provided in Table 5.11.

Here, again, at the 0.05 significance level, only factor A is statistically significant. This result means that factor A with value 2 generates the best outcome. The difference observed in the product is not related to the values of B or the interaction between A and B.

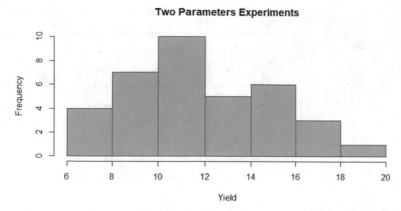

**Two Parameters Experiments**

Fig. 5.9 Example 5.7: two parameters experiment

**Table 5.11**  Example 5.7: ANOVA evaluation using R language library AICcmodavg

|                           | Df | Sum Sq | Mean Sq | F value | Pr(>F) |
| ------------------------- | -- | ------ | ------- | ------- | ------ |
| Parameter A               | 2  | 60.39  | 30.194  | 3.514   | 0.044  |
| Parameter B               | 2  | 22.39  | 11.194  | 1.303   | 0.288  |
| Parameter A: Parameter B  | 4  | 36.44  | 9.111   | 1.060   | 0.395  |
| Residuals                 | 27 | 232.00 | 8.593   |         |        |

**Example 5.8: 3-Factors, 2-Levels Full Factorial Designs**  This experiment's objective is to identify three elements in an amplifier, where $W$ is the width of the microstrip lines, $R$ is the resistor, and $C$ is the capacitor, which is most important in magnifying the amplifier or amplifier gain.

For this experiment, we will change the three variables between two levels, $-1$ and $+1$, which represent the following values:

- W: $-1$ represents $-20\%$ variations and $+1$ represents $+20\%$ variation with respect to the nominal or base value.
- R: $-1$ represents $-5\%$ variation and $+1$ represents $+5\%$ variation with respect to the nominal or base value.
- C: $-1$ represent $-10\%$ variation and $+1$ represents $+10\%$ variation with respect to the nominal or base value.

A full factorial design was prepared, and the eight (8) results of the experiments are shown in Table 5.12 in the Gain column.

This design was analysed using the *RcmdrPlugin.DoE* library in R language.

The experiment's outcomes can be represented using the dot plot in Fig. 5.10, where we can observe that data group into clusters of equal size, around 13.0 and 14.5.

**Table 5.12** Example 5.8:
3-factors, 2-levels full
factorial design

| W  | R  | C  | GAIN  |
|----|----|----|-------|
| −1 | −1 | −1 | 12.85 |
| +1 | −1 | −1 | 13.01 |
| −1 | +1 | −1 | 14.52 |
| +1 | +1 | −1 | 14.71 |
| −1 | −1 | +1 | 12.93 |
| +1 | −1 | +1 | 13.09 |
| −1 | +1 | +1 | 14.61 |
| +1 | +1 | +1 | 14.81 |

**Fig. 5.10** Example 5.8: a
dot plot of the Electric design

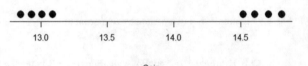

Gain

The dot diagram shows the general location or central tendency of the observations and their spread; it is useful when the data points are lower than 20. When more than 20 observations are available, it is better to use a histogram.

The histogram shows the central tendency, spread, and general shape of the data distribution; it is built by dividing the horizontal axis into bins and drawing a rectangle over each bin with an area proportional to the number of observations in the corresponding bin.

The plot called *strip chart* shows a direct correlation between the input parameters' values and the outputs. Figures 5.11, 5.12, and 5.13 show the correlation of each separate variable.

The half-normal plot for assessing 'effect significance' is shown in Fig. 5.14.

This plot shows that, within 90% confidence (a significance level of 0.1), the most critical variable is the resistor (R) value, followed by the width of lines (W) and the capacitor (C). The half-normal plot shows that the interaction effect of the resistor and width of the line (AB) is the fourth variable in terms of relevance.

The analysis of this design continues by using a linear model of the main effects and interactions. This linear model (see Table 5.13) identifies the same order of relevance for the three factors and the interaction factors.

Figure 5.15 shows the main effects; it indicates that the resistor is the most critical factor, as concluded before. Far from this, the width of lines and the capacitors follow in the level of importance.

Figure 5.16 shows the interaction effects between the width of lines, resistors, and capacitors; the more significant interactions are the width of lines/resistors and resistors/capacitors.

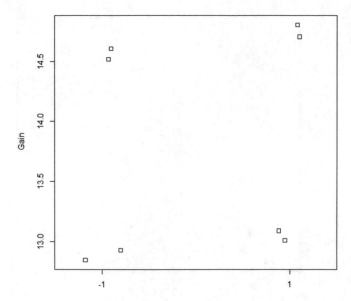

**Fig. 5.11** Example 5.8: the correlation between the variable width of lines and the outcomes

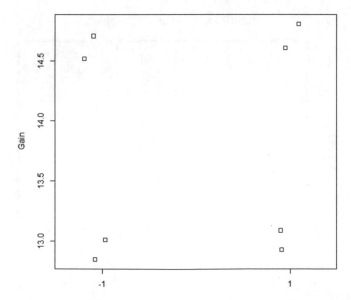

**Fig. 5.12** Example 5.8: the correlation between the variable capacitors and the outcomes

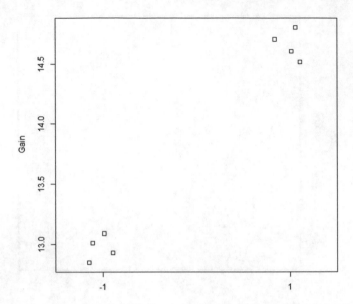

**Fig. 5.13** Example 5.8: the correlation between the variable resistors and the outcomes

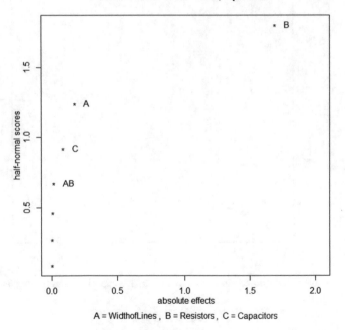

**Fig. 5.14** Example 5.8: half normal plot for amplifier gain

**Table 5.13** Example 5.8: a linear model for the electrical design

|                              | Estimate | Std. error | t value | Pr(>|t|)  |
|------------------------------|----------|------------|---------|-----------|
| (Intercept)                  | 13.81625 | 0.00125    | 11,053  | 0.0000576 |
| Width of Lines1              | 0.08875  | 0.00125    | 71      | 0.00897   |
| Resistors1                   | 0.84625  | 0.00125    | 677     | 0.00094   |
| Capacitors1                  | 0.04375  | 0.00125    | 35      | 0.01818   |
| Width of Lines1: Resistors1  | 0.00875  | 0.00125    | 7       | 0.09033   |
| Width of Lines1: Capacitors1 | 0.00125  | 0.00125    | 1       | 0.50000   |
| Resistors1: Capacitors1      | 0.00375  | 0.00125    | 3       | 0.20483   |

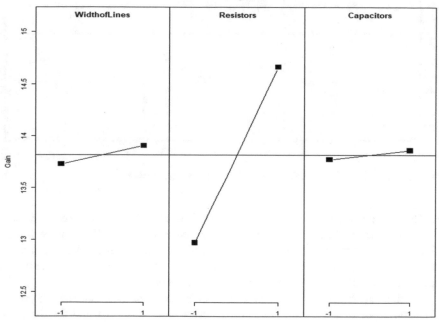

**Fig. 5.15** Example 5.8: main effects plot for the electric design

Interaction effects can also be displayed using the linear model; they have the advantage of showing confidence bands in this R library. The disadvantage is the different scales displayed (Fig. 5.17).

**Example 5.9: Cake-Mix Design—3-Factors, 2-Levels Full Factorial with 3-Centre Points Design** The cake-mix design refers to an example of an industrial plant for producing a cake mix. There are instructions and recommendations on preparing the cake and the different product components in the box.

**Interaction plot matrix for Gain**

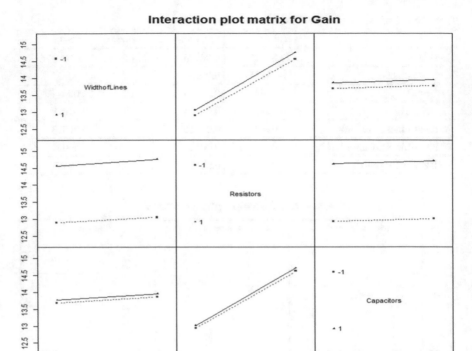

**Fig. 5.16**   Example 5.8: interaction effects plot for the electric design

The cake mix comprises many components. This example will focus on only three components (ingredients): *Flour*, *Shortening,* and *Egg powder*. These experiments aimed to find the optimal combination of ingredients that produces the best flavour. A jury consisting of three experts was selected to judge the quality of the recipe under the variable 'taste': the higher the value, the better the taste.

For this experimental design, we use a factorial design with three centre points. The centre points correspond with the analysts' values are optimal for the three ingredients to produce the best flavour. The selection of three centre points was made to have a more accurate reference value. For the remaining trials, we have $2^k$ combinations which, in this case, added eight (8) trials. The matrix design and the results of each trial are shown in Table 5.14.

Figure 5.18 shows the schematic of the cake-mix experiment.

To analyse the results of these experiments, we first used the half-normal plot, which uses one point per factor and interaction factor. In Example 5.9, we have seven points in the half-normal plot between factors and interaction factors, as shown in Fig. 5.19.

From the Normal plot, at a significance level of 0.1, we conclude that the more critical factors are the interactions between *Shortening: EggPowder*, *EggPowder* and

**WidthofLines\*Resistors effect plot   WidthofLines\*Capacitors effect plot**

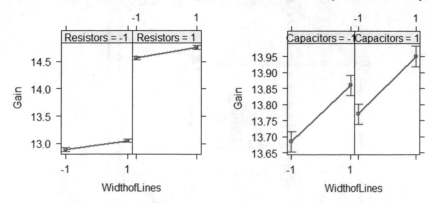

**Resistors\*Capacitors effect plot**

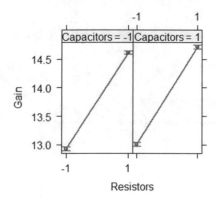

**Fig. 5.17** Example 5.8: interaction effects from a linear model

**Table 5.14** Example 5.9: cake-mix experimental design and outcomes

| Trial | Flour | Shortening | Egg powder | Taste |
|---|---|---|---|---|
| 1 | 200 | 50 | 50 | 3.52 |
| 2 | 400 | 50 | 50 | 3.66 |
| 3 | 200 | 100 | 50 | 4.74 |
| 4 | 400 | 100 | 50 | 5.20 |
| 5 | 200 | 50 | 100 | 5.38 |
| 6 | 400 | 50 | 100 | 5.90 |
| 7 | 200 | 100 | 100 | 4.36 |
| 8 | 400 | 100 | 100 | 4.86 |
| 9 | 300 | 75 | 75 | 4.73 |
| 10 | 300 | 75 | 75 | 4.61 |
| 11 | 300 | 75 | 75 | 4.68 |

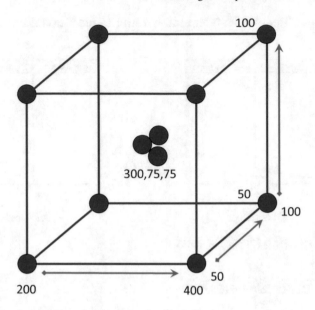

**Fig. 5.18** Example 5.9: cake-mix experiment's schematic

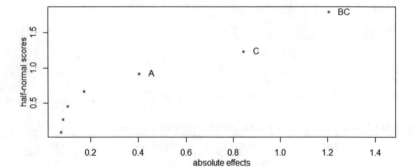

**Fig. 5.19** Example 5.9: half normal plot for cake mix experiment

*Flour* effects; these three effects (one interaction and two main effects) are distinctly apart from the other four effects.

The effects plots, shown in Fig. 5.20, illustrate the changes in the response due to changes in the effects.

In the main effect plots, it can be seen that the most crucial main effect for EggPowder and the less important is that for Shortening. It can also be seen that, for all three factors, the higher the value, the higher the taste (outcome).

Similarly, the interaction effects are displayed in Fig. 5.21.

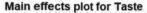

**Main effects plot for Taste**

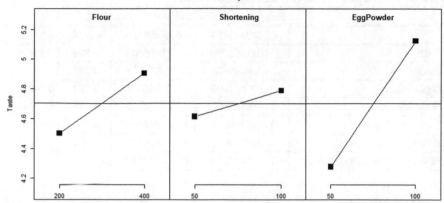

**Fig. 5.20**  Example 5.9: main effects plots for the cake-mix

**Interaction plot matrix for Taste**

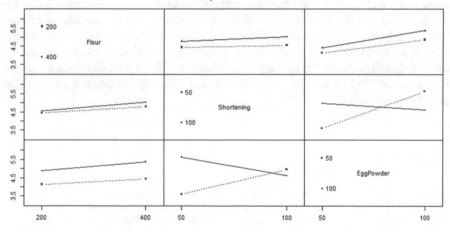

**Fig. 5.21**  Example 5.9: interaction effects plots for the cake-mix

The linear model for this example is shown in Table 5.15.

## 5.8  Summary

In this chapter, we discuss the rationale of experimental design and the reasons for
its importance. We review the historical development of DOE and the several uses
of this technique. A detailed description of DOE and the terminology is presented,

**Table 5.15**  Example 5.5: cake mix linear regression model

|  | Estimate | Std. error | t value | Pr(>|t|) |
|---|---|---|---|---|
|  | −0.9079545455 | 0.6164494862 | −1.473 | 0.23720 |
| Flour | −0.0045000000 | 0.0019496503 | −2.308 | 0.10421 |
| Shortening | 0.0560000000 | 0.0077986013 | 7.181 | 0.00556 |
| EggPowder | 0.0676000000 | 0.0077986013 | 8.668 | 0.00323 |
| Flour: Shortening | 0.0000660000 | 0.0000246613 | 2.676 | 0.07529 |
| Flour: EggPowder | 0.0000720000 | 0.0000246613 | 2.920 | 0.06152 |
| Shortening: EggPowder | −0.0007600000 | 0.0000986454 | −7.704 | 0.00454 |
| Flour: Shortening: EggPowder | −0.0000006800 | 0.0000003119 | −2.180 | 0.11736 |

including several simple applications; the contrast between DOE and other experimental methods is discussed. DOE workflow and its principles are explained in detail. DOE is implemented for screening purposes, and, in that sense, we focus on the full factorial design, and several examples are discussed. The relation between uncertainty and DOE is also presented. Experimental design can be used during the project's assessment of the value of information for assessing the impact, uncertainties, and relative importance of variables in the project's outcomes. In a project where several uncertainties are present, DOE is specially tailored to assess the impact of the variables' uncertainty in the project outcome. In this way, DOE guides the value of the information process by suggesting which data acquisition action should be preferably taken.

# References

Amudo, C., Graf, T., Dandekar, R., & Randle, J. (2008). The pains and gains of experimental design and response surface applications in reservoir simulation studies. *Society of Petroleum Engineers*. In *Proceeding of the 2009 Reservoir Simulation Symposium*, The Woodlands, Texas, USA, 2–4 February 2009. Paper SPE-118709.

Box, G., & Hunter, J. (1957). Multifactor experimental design for exploring response surfaces. *Annals of Mathematical Statistics, 28*, 195–242.

Durakovic, B. (2017). Design of experiments application, concepts, examples: State of the art. *Periodicals of Engineering and Natural Sciences, 5*(3), 421–439.

Fisher, F. (1935). *The design of experiments*. Oliver and Boyd Ltd.

Moeinikia, F., & Alizadeh, N. (2012). Experimental design in reservoir simulation: an integrated solution for uncertainty analysis, a case study. *Journal of Petroleum Exploration and Production Technology, 2*, 75–83.

Montgomery, D. (2013). *Design and analysis of experiments* (8th ed.). Wiley.

Oehlert, G. (2010). *A first course in design and analysis of experiments*. Ed. W.H. Freeman and Company.

Tanco, M. (2008). *Metodologia para la aplicación del diseño de experimentos (DOE) en la industria (Doctoral tesis)*. Escuela Superior de Ingenieros de la Universidad de Navarra.

Telford, J. (2007). A Brief introduction to design of experiments. *John Hopkins APL Technical Digest, 27*(3), 224–232.

White, Ch., & Royer, S. (2003). Experimental design as a framework for reservoir studies. *Society of Petroleum Engineers.* In *Proceeding of the SPE Reservoir Simulation Symposium*, Houston, Texas, USA, 3–5 February 2003. Paper SPE-79676.

# Chapter 6
# Fuzzy Logic

**Objective**

In this book, fuzzy logic is used to incorporate the uncertainty due to vagueness into the methodology of the value of information to account for the vagueness of the data. Crisp versus fuzzy data is discussed. Membership functions, fuzzification, defuzzification, fuzzy rules and fuzzy inference systems are discussed in detail with an example.

## 6.1   Introduction to Fuzzy Logic

The classical or Boolean logic uses sharp distinctions to define a class of elements: a given element belongs entirely to a set or does not belong to it at all. E.g., if the definition of a tall person is someone more than 1.80 m and less than that is a medium person, then Tom, with a height of 1.81 m, is tall while John, at 1.78 m, is medium; however, is that a fair description? Does it reflect reality? Fuzzy logic is designed to deal with these situations, providing a solution that closely follows human logic. In that sense, fuzzy logic reflects what people think, and it gives the equipment to create more intelligent systems. Fuzzy logic is a superset of Boolean logic dealing with the concept of partial truth. In contrast, classical logic assumes that everything can be expressed in binary terms, 0 and 1, black and white, while fuzzy logic assumes that there are degrees of truth.

Logic is the branch of knowledge that studies the structure and principles of correct reasoning, establishing the principles for valid deductive arguments. Logic deals with the two opposite concepts of truth and falsehood.

The idea underlying the dichotomy between truth and falsehood is supported by the validity of two laws of classical logic, which will be explained in Sect. 6.2, Eqs. (6.30) and (6.31):

I.   *Law of the excluded middle*: every proposition is true or false, and there is no other possibility.
II.  *Law of no contradiction*: no statement is true and false at the same time.

These laws underpin the classical set theory. However, they are not the only logical option. As will be discussed in Sect. 6.3, they are not valid in fuzzy set theory.

The origins of the fuzzy logic date back to the Greek philosopher Plato (428–347 B.C.), who considered different degrees between truth and falsehood. This simple concept can be translated into many life situations: several degrees exist between black and white.

David Hume (1711–1776), the Scottish philosopher, economist, and historian, proposed common-sense logic supported by the reasoning acquired by our lives' experiences. At the same time, Immanuel Kant (1724–1804), the German philosopher, argued that only a mathematician could provide clear and precise definitions for logical reasoning.

Charles Sanders Peirce (1839–1914), an American philosopher, logician, scientist and mathematician, and the British mathematician, philosopher, and logician Bertrand Russell (1872–1970), reported that classical logic led to contradictions, and they studied the problem of vagueness. Jan Lukasiewicz (1878–1956), the Polish logician and philosopher, first proposed in 1920 the three-valued or trivalent logic; Lukasiewicz, in addition to true and false, accepts an indeterminate truth value of membership of 0.5. Subsequently, Lukasiewicz introduces the idea of possibility theory by extending the range of valid values of all real numbers between 0 and 1; these numbers are not a probability but a possibility that an element belongs to a set; possibility theory is an inexact reasoning technique.

Both Bertrand Russell's formulation of the paradox of all sets that do not contain themselves and the German theoretical physicist Werner Heisenberg (1901–1976) with the principle of uncertainty of quantum mechanics showed some of the contradictions of the classical logic.

The theory of "vague sets" or "fuzzy sets" was proposed by the British-American physicist and philosopher Max Black (1909–1988), who analysed the problem of "vagueness" in his paper "Vagueness: an exercise of logical analysis" (1937). Black discussed the vagueness of terms or symbols and proposed a consistency profile or curve to analyse vagueness, similar to what is now known as *membership functions*. Black suggested and discussed the example called *chair similarity*: next to a chair, we locate another chair that lacks a small portion of it and can be described as "less-chair" than the first one; then another chair is located next to the second one, which lacks yet another small portion and as such is "less-chair" than the second. This process continues, and at the end, we get a log. The first chair has a membership 1 in the set of chairs, but the "membership number" is reduced from chair to chair until we get the log. Black asked himself: when is the chair no longer a chair? The chair as a concept does not establish a clear boundary of what is and what is not a chair.

This same situation occurs with an apple; we can say that one specific apple belongs to the apple set (because it is an apple, it belongs in full to the set of all apples). If we create two additional sets: a set of ripe apples and another set of unripe apples, some of the apples belong in full to one set or the other, but some belong partially to one set depending on the level of ripeness.

In crisp sets, an element is either included in the set or is not. In fuzzy logic, an element is included with a degree of truth from 0 to 1; fuzzy logic models allow an element to be categorised in more than one set with different levels of truth or confidence. Fuzzy logic recognises the lack of knowledge or absence of precise data.

These are the foundations for the theory presented by the mathematician, electrical engineer, computer scientist, artificial intelligence researcher and professor Lotfi Zadeh in 1965: fuzzy sets. A fuzzy set is a collection of objects that might belong to a set with a degree varying from 1 for full belongingness to 0 for full non-belongingness. With this idea in his mind, Zadeh intended to create a formalism to efficiently handle the imprecision in human reasoning. Zadeh used a membership function concept, assigning each element a number from the unit interval to indicate the intensity of belongingness. Further, in 1973 Zadeh published another capital paper introducing the concept of "linguistic variables", which is essential to understand the development of fuzzy inference systems explained in Sect. 6.9 of this book.

In the fuzzy theory, the classical concepts of truth and falsehood are just the extreme cases of the broader idea of vagueness or fuzziness. In the fuzzy concept, a proposition can be partially true and partially false simultaneously; in that sense, a person is not tall or short but partially tall and partially short. So, everything has different degrees of belonging (membership) to the sets that describe the discourse universe. Fuzzy logic is the theory of fuzzy sets that calibrate the vagueness.

The fuzzy set theory's fundamental idea is that an element belongs to a fuzzy set with a certain degree of membership; so, a proposition may be partially true or partially false to any degree; this degree of belonging is described with a number from the real set [0, 1].

## 6.2  Classical Set Theory

A *universe of discourse*, $X$, is a collection of objects all with the same characteristics; examples of elements of the universe are the real numbers between 0 and 1, cars with a speed limit within a range of 0–240 km/h, the temperature levels of an operating pump with a range between 0 and 100 °C, etc. The individual elements of the universe of discourse are denoted as $x$.

If the universe's elements are continuous, the set defined on the universe is made of continuous elements; otherwise, they are discrete elements.

The cardinal number of a universe $X$, $n_x$ is the total number of elements in the universe. If the universe has a discrete and finite number of elements, it has a finite cardinal number; if the universe has continuous elements, its number is infinite and has an infinite cardinality.

Collections of elements in a universe are called sets. The null set Ø contains no elements, and the whole set, $X$, is the set containing all the elements of the universe. The power set, $P(X)$ is the set containing all possible sets of $X$.

Let $X$ be a classical or crisp set and $x$ an element; $x$ either belongs to $X$ or does not belong to $X$. So, elements are identified by 1 if they belong to $X$ or 0 if not. This assignment is called the principle of dichotomy.

The classical set theory was developed by the German mathematician Georg Cantor (1845–1918). It describes the interaction of crisp sets, which is called operations, described in the next section.

Operations in classical sets

The operations in classical set theory are:

(i)   *Union*: the union between two sets $A$ and $B$, $A \cup B$, is the set containing all the elements of $A$ and all the elements of $B$ (independently if they belong simultaneously to the two sets). This operation is called the logical "or". In the set theory,

$$A \cup B = \{x \,|\, x \in A \text{ or } x \in B\} \tag{6.1}$$

The Venn diagram of the union operation is represented in Fig. 6.1.

The union operation is the reverse of the intersection, as we will explain next.

**Example 6.1: Union in Classical Sets**   Assume we define the universe of discourse $= \{1, 2, 3, 4, 5, 6, 7, 8, 9, 10\}$, integer numbers from 1 to 10, and two crisp sets $A$, $B$, as in Eqs. (6.2) and (6.3):

**Fig. 6.1** Union in classical set theory

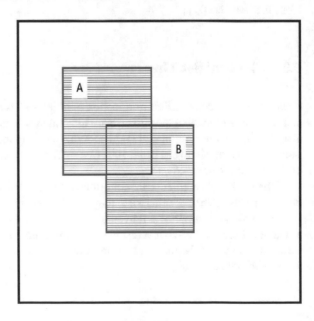

$$A = \{1, 2, 3, 4, 5\} \tag{6.2}$$

$$B = \{2, 4, 6, 8\} \tag{6.3}$$

The union of $A$ and $B$ is defined as

$$A \cup B = \{1, 2, 3, 4, 5, 6, 8\} \tag{6.4}$$

(ii)   *Intersection*: the intersection between two sets $A$ and $B$, $A \cap B$, is the set containing all the elements of the universe that belong simultaneously to $A$ and $B$. This operation is called the "and" operator. In the set theory,

$$A \cap B = \{x \,|\, x \in A \; and \; x \in B\} \tag{6.5}$$

The Venn diagram of the intersection property is shown in Fig. 6.2.

**Example 6.2: The Intersection in Classical Sets** Assuming the sameets discussed before, $A$ and $B$, as in Eqs. (6.6) and (6.7),

$$A = \{1, 2, 3, 4, 5\} \tag{6.6}$$

$$B = \{2, 4, 6, 8\} \tag{6.7}$$

The intersection of $A$ and $B$ is

**Fig. 6.2** The intersection in classical set theory

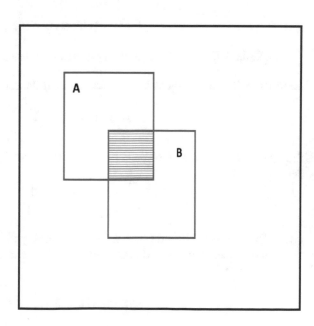

**Fig. 6.3** *Complement in classical set theory*

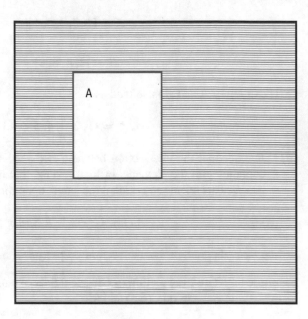

$$A \cap B = \{2, 4\} \tag{6.8}$$

(iii)   *Complement*: the complement of set $A$, $\overline{A}$ is the set of all the elements of the universe that do not belong to the set $A$. The complement of a set is the opposite of this set, represented in set theory by Eq. (6.9):

$$\overline{A} = \{x \,|\, x \notin A, x \in X\} \tag{6.9}$$

The Venn diagram of the complement operation is shown in Fig. 6.3.

**Example 6.3: Complement in Classical Sets**  If the set $A$ is given by Eq. (6.10),

$$A = \{1, 2, 3, 4, 5\} \tag{6.10}$$

then the complement of $A$ is given by Eq. (6.11):

$$\overline{A} = \{6, 7, 8, 9, 10\} \tag{6.11}$$

(iv)   *Difference*: the difference of a set $A$ concerning the set $B$, $A|B$, is the set of all the elements of $A$ that do not belong to $B$. The difference operation is shown in Eq. (6.12):

$$A|B = \{x \,|\, x \in A \text{ and } x \notin B\} \tag{6.12}$$

**Fig. 6.4** The difference in classical set theory

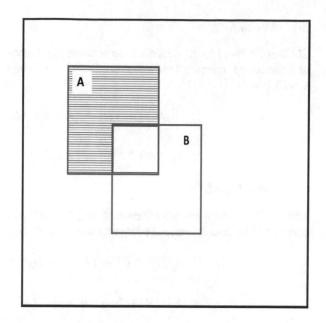

Using the Venn diagram, the difference operation is shown in Fig. 6.4.

**Example 6.4: The Difference in Classical Sets** Given two sets $A$ and $B$, as per Eqs. (6.13) and (6.14),

$$A = \{1, 2, 3, 4, 5\} \tag{6.13}$$

$$B = \{2, 4, 6, 8\} \tag{6.14}$$

Their set difference is shown in Eq. (6.15):

$$A|B = \{1, 3, 5\} \tag{6.15}$$

Properties of classical sets

The properties of classical sets describe the manipulations that can be made with the sets.

(i)  *Commutativity*

The commutative property is described with Eqs. (6.16) and (6.17):

$$A \cup B = B \cup A \tag{6.16}$$

$$A \cap B = B \cap A \tag{6.17}$$

(ii)   *Associativity*

The associative property describes the iterative application in different order of union or intersection operations between three sets, $A$, $B$, and $C$, as shown in Eqs. (6.18) and (6.19):

$$A \cup (B \cup C) = (A \cup B) \cup C \tag{6.18}$$

$$A \cap (B \cap C) = (A \cap B) \cap C \tag{6.19}$$

(iii)   *Distributivity*

The distributive property describes the iterative application in different order of union and/or intersection operations between three sets, $A$, $B$, and $C$ :

$$A \cup (B \cap C) = (A \cup B) \cap (A \cup C) \tag{6.20}$$

$$A \cap (B \cup C) = (A \cap B) \cup (A \cap C) \tag{6.21}$$

(iv)   *Idempotency*

The idempotent property describes the way a set $A$ relates with itself, with the null set and with the universe of discourse set, as shown in Eqs. (6.22) and (6.23):

$$A \cup A = A \tag{6.22}$$

$$A \cap A = A \tag{6.23}$$

(v)   *Identity*

The identity property shows the relations of the set $A$ with the universe and the null set as shown in Eqs. (6.24)–(6.27):

$$A \cup \emptyset = A \tag{6.24}$$

$$A \cap \emptyset = \emptyset \tag{6.25}$$

$$A \cup X = X \tag{6.26}$$

$$A \cap X = A \tag{6.27}$$

(vi)   *Transitivity*

The transitivity property represents a logical relation between three sets based on the inclusion relation, as described in Eq. (6.28):

$$If \ A \subseteq B \ and \ B \subseteq C, then \ A \subseteq C \tag{6.28}$$

(vii)   *Involution*

Similar to the transitivity property, the involution describes a logical consequence of the complementary function, Eq. (6.29):

$$\overline{\overline{A}} = A \tag{6.29}$$

(viii)   *The excluded middle law*

This law is valid for the classical (crisp) set, but it is not valid for fuzzy sets, as discussed later in this chapter. This law for the crisp set has two parts, Eqs. (6.30) and (6.31):

(a)   *Law of the excluded middle*:

$$A \cup \overline{A} = X \tag{6.30}$$

This law says that whatever is not part of the set $A$ is part of its complement $\overline{A}$.

(b)   *Law of no contradiction*:

$$A \cap \overline{A} = \emptyset \tag{6.31}$$

This law says that there are no common elements between a set $A$ and its complement $\overline{A}$.

(ix)   *De Morgan's principles*

De Morgan's principles have two parts, as shown in Eqs. (6.32) and (6.33):

$$(a) \quad \overline{A \cap B} = \overline{A} \cup \overline{B} \tag{6.32}$$

$$(b) \quad \overline{A \cup B} = \overline{A} \cap \overline{B} \tag{6.33}$$

De Morgan's principle says that the complement of the union or intersection of two sets is equal to the intersection of the union or intersections of the sets' complements.

The relations described by De Morgan's principle can be extended to any number of sets.

Figures 6.5 and 6.6 show the representation of the two parts of De Morgan's principle.

**Fig. 6.5**  De Morgan's
$\overline{A \cap B}$

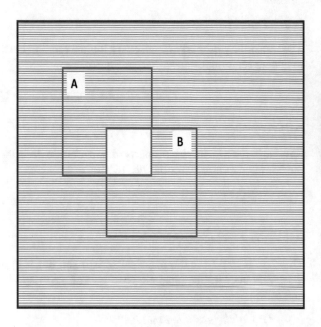

**Fig. 6.6**  De Morgan's
$\overline{A \cup B}$

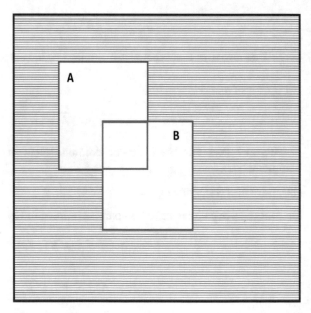

Sets can be defined using one of three methods:

(i)   The list method: naming all its members.

$$A = \{a_1, a_2, \ldots, a_n\} \qquad (6.34)$$

(ii)    The rule method: specifying the property satisfied by its members.

$$A = \{x \mid P(x)\} \tag{6.35}$$

(iii)   Characteristic function method: defining a function,

$$\chi_A(x) = \begin{cases} 1 & for\ x \in A \\ 0 & for\ x \notin A \end{cases} \tag{6.36}$$

Elements of one universe of discourse can be mapped on elements of another universe of discourse. When an element $x$ contained in a set $X$ is mapped in an element $y$ contained in the set $Y$, this mapping is represented by Eq. (6.37):

$$f : X \rightarrow Y \tag{6.37}$$

The concepts, operations, and relations defined in set theory can be implemented in practical applications using mappings between set elements and real numbers; indeed, elements are mapped from one universe of discourse into the Real Numbers universe using functions.

Suppose we have two universes of discourse, $X$ and $Y$, with elements $x$ and $y$ respectively. A critical mapping is the so-called *characteristic function*, $\chi_A$, defined by Eq. (6.38):

$$\chi_A(x) : X \rightarrow \{0, 1\} \tag{6.38}$$

where

$$\chi_A = \begin{cases} 1, & if\ x \in A \\ 0, & if\ x \notin A \end{cases} \tag{6.39}$$

The characteristic function maps a set of elements $x$ in $A$ (a crisp set to differentiate from the fuzzy set that we will discuss later) to the real numbers 1 or 0.

$\chi_A$, the characteristic function represents the "*membership*" of the elements in $A$: if they belong to $A$ this function assigns the number 1, and if they do not belong to $A$ this function gives the value 0. In this sense, the mapping is between set $A$ which is a set of the universe $X$ and the set made of two real numbers, 1 and 0, a set of the universe $Y$.

The mapping function is used to execute operations between the elements of sets. Suppose we have two sets $A$ and $B$ defined on the same universe $X$; the operations previously defined using diagrams on the classical sets can be implemented between the elements of sets, using the characteristic function as described below.

Union:

$$A \cup B \rightarrow \chi_{A \cup B}(x) = \chi_A(x) \vee \chi_B(x) = \max(\chi_A(x), \chi_B(x)) \tag{6.40}$$

Intersection:

$$A \cap B \rightarrow \chi_{A \cup B}(x) = \chi_A(x) \wedge \chi_B(x) = \min(\chi_A(x), \chi_B(x)) \qquad (6.41)$$

Complement:

$$\overline{A} \rightarrow \chi_{\overline{A}}(x) = 1 - \chi_A(x) \qquad (6.42)$$

Containment:

$$A \subseteq B \rightarrow \chi_A(x) \leq \chi_B(x) \qquad (6.43)$$

where the symbols $\vee$ and $\wedge$ are the maximum and minimum operators.

## 6.3 Fuzzy Set Theory

In human communication, we use many vague expressions, e.g., we can say that the speed at which a car moves on a highway is "low", "medium", or "fast", and humans easily understand these categories. However, when more accuracy is needed, we can say:

i.    Low speed: a car that moves between 0 and 40 km/h.
ii.   Medium speed: a vehicle that drives between 40 and 80 km/h.
iii.  Fast speed: a vehicle that moves faster than 80 km/h.

Nevertheless, what happens if a car's speed is 81 km/h? Is it a fast or medium speed? Or if the speed is 39 km/h, is it low or medium speed?

If we think like a computer, following the Boolean or classical set theory, there is no doubt that the first case corresponds to a fast speed and the second case is a low speed. However, 39 km/h is partially in the two categories between low and medium in human logic. Clearly, 0 km/h is entirely in the low-speed category, but that is not the case for 39 km/h; similarly, 81 km/h, in the human logic, belongs partially to the two-speed categories, medium and fast, although in the classical set theory it belongs entirely to the fast speed category and does not belong to the medium category. In this sense, classical logic is a crisp logic with total belonging; however, the real world seems less strict.

Fuzzy logic is a multivalued logic that represents uncertainty and vagueness mathematically. Any problem can be understood as a set of input variables combined to obtain a set of outcome variables; fuzzy logic establishes that mapping, following criteria of meaning, and not precision criteria.

Even though fuzziness and probability are forms of uncertainty, they are different: fuzziness describes the ambiguity with which a piece of data belongs to one set. In contrast, probability represents the likelihood of a piece of data belonging to a set.

Fuzzy logic defines the degree of belonging of elements to a set, using values from zero (0—not belonging) to one (1—fully belonging); in that sense, binary logic is a particular case of fuzzy logic.

In fuzzy logic, the values are associated with fuzzy sets by a fuzzification process; the fuzzified values can be manipulated using linguistic rules. The output can remain fuzzy or be defuzzified to obtain a crisp value. Fuzzy sets theory develops a series of concepts to systematically handle the imprecision occurring when the limits between classes of objects are not clearly defined.

Fuzzy theory generalises the concept of the characteristic function Eq. (6.36) to describe the degree of partial belonging through the membership function $\mu_{\tilde{M}}$, such that, for a fuzzy set $\tilde{M}$,

$$\mu_{\tilde{M}} = X \rightarrow [0, 1] \tag{6.44}$$

where,
$\mu_{\tilde{M}}(x) = 1$ if $x$ is fully in $\tilde{M}$;
$\mu_{\tilde{M}}(x) = 0$ if $x$ is not in $\tilde{M}$;
$0 < \mu_{\tilde{M}}(x) < 1$ if $x$ is partially in $\tilde{M}$.

Fuzzy sets are denoted by a set symbol with a tilde overstrike, $\tilde{M}$.

The larger the value that the membership function assigns to an element in a set, the higher the degree of belonging. Elements of a fuzzy set can also be members of other sets in the same universe. As discussed in the previous section, in the classical set theory, an element belongs "totally" to a set, or it does not belong at all to the set; this is the binomial logic, 0 and 1 or yes and no. In that sense, there is a crisp meaning of belonging.

In fuzzy theory, elements can belong "partially" to one set, represented by a function that describes such ambiguity. Different elements have different membership levels in a fuzzy set; however, that is not true in classical sets.

A fuzzy set $\tilde{A}$ of discrete elements is described by its elements as per Eq. (6.45):

$$\tilde{A} = \left\{ \frac{\mu_{\tilde{A}}(x_1)}{x_1} + \frac{\mu_{\tilde{A}}(x_1)}{x_2} + \cdots \right\} = \left\{ \sum_i \frac{\mu_{\tilde{A}}(x_i)}{x_i} \right\} \tag{6.45}$$

In this Equation, the horizontal bar is not the sign of division but a separator. The sign "+" is not the algebraic addition but just an aggregation; in the same way, the summation at the right-hand side is not the algebraic summation but an aggregation.

Extending these meanings to the continuous case, a continuous fuzzy set $\tilde{A}$ is given by Eq. (6.46):

$$\tilde{A} = \left\{ int \frac{\mu_{\tilde{A}}(x)}{x} \right\} \tag{6.46}$$

Operations on fuzzy sets

Consider two fuzzy sets $\tilde{A}$ and $\tilde{B}$, of a shared universe $X$; for a given element $x$ of the universe, the union, intersection, and complement operations are defined as per Eqs. (6.47), (6.51) and (6.55):

(i)    Union: the union operation in fuzzy sets is given by

$$\mu_{\tilde{A} \cup \tilde{B}}(x) = \mu_{\tilde{A}}(x) \vee \mu_{\tilde{B}}(x) \tag{6.47}$$

The union operation for the fuzzy set gives the largest membership value for the common elements.

**Example 6.5: Union in Fuzzy Sets**  Let us assume we have the fuzzy sets $\tilde{A}$ and $\tilde{B}$ :

$$\tilde{A} = \left\{ \frac{0.00}{10}, \frac{0.20}{30}, \frac{0.70}{60}, \frac{1.00}{90} \right\} \tag{6.48}$$

$$\tilde{B} = \left\{ \frac{0.10}{10}, \frac{0.25}{40}, \frac{0.35}{50}, \frac{0.50}{90} \right\} \tag{6.49}$$

then their union is given by Eq. (6.50):

$$\tilde{A} \cup \tilde{B} = \left\{ \frac{0.10}{10}, \frac{0.20}{30}, \frac{0.25}{40}, \frac{0.35}{50}, \frac{0.70}{60}, \frac{1.00}{90} \right\} \tag{6.50}$$

As in a crisp set, the union operation for fuzzy sets is the reverse of the intersection operation.

(ii)    Intersection: the intersection between two sets contains the elements that belong simultaneously to the two sets, but each element can have a different degree of membership in the two sets for fuzzy sets. The intersection degree is the lowest degree of membership in both sets of each element:

$$\mu_{\tilde{A} \cap \tilde{B}}(x) = \mu_{\tilde{A}}(x) \wedge \mu_{\tilde{B}}(x) \tag{6.51}$$

**Example 6.6: The Intersection in Fuzzy Sets**  Let us assume the fuzzy sets $\tilde{A}$ and $\tilde{B}$ are described by Eqs. (6.52) and (6.53):

$$\tilde{A} = \left\{ \frac{0.00}{10}, \frac{0.20}{30}, \frac{0.70}{60}, \frac{1.00}{90} \right\} \tag{6.52}$$

$$\tilde{B} = \left\{ \frac{0.10}{10}, \frac{0.25}{40}, \frac{0.35}{50}, \frac{0.50}{90} \right\} \tag{6.53}$$

The intersection of $\tilde{A}$ and $\tilde{B}$ is calculated as shown in Eq. (6.54):

$$\tilde{A} \cap \tilde{B} = \left\{ \frac{0.00}{10}, \frac{0.50}{90} \right\} \qquad (6.54)$$

(iii)   Complement: the complement of a set is the opposite of this set:

$$\mu_{\overline{\tilde{A}}}(x) = 1 - \mu_{\tilde{A}}(x) \qquad (6.55)$$

**Example 6.7: Complement in Fuzzy Sets**  Let us assume the fuzzy set $\tilde{A}$:

$$\tilde{A} = \left\{ \frac{0.00}{10}, \frac{0.20}{30}, \frac{0.70}{60}, \frac{1.00}{90} \right\} \qquad (6.56)$$

The complement of $\tilde{A}$, $\overline{\tilde{A}}$ is

$$\overline{\tilde{A}} = \left\{ \frac{1.00}{10}, \frac{0.80}{30}, \frac{0.30}{60}, \frac{0.00}{90} \right\} \qquad (6.57)$$

As obtained in the classical set theory, assuming that $X$ is the universe and $\emptyset$ is the null set, for the fuzzy sets:
  For all

$$x \in X, \mu_{\emptyset}(x) = 0 \qquad (6.58)$$

For all

$$x \in X, \mu_X(x) = 1 \qquad (6.59)$$

Properties of fuzzy sets

The properties previously defined for the classical set can be extended, naturally, to the fuzzy set.

(i)   *Commutativity*

The commutative property between two fuzzy sets $\tilde{A}$ and $\tilde{B}$ is given by Eqs. (6.60) and (6.61):

$$\tilde{A} \cup \tilde{B} = \tilde{B} \cup \tilde{A} \qquad (6.60)$$

$$\tilde{A} \cap \tilde{B} = \tilde{B} \cap \tilde{A} \qquad (6.61)$$

(ii)   *Associativity*

Given three fuzzy sets, $\tilde{A}$, $\tilde{B}$ and $\tilde{C}$, the associative property is described by Eqs. (6.62) and (6.63):

$$\tilde{A} \cup \left(\tilde{B} \cup \tilde{C}\right) = \left(\tilde{A} \cup \tilde{B}\right) \cup \tilde{C} \tag{6.62}$$

$$\tilde{A} \cap \left(\tilde{B} \cap \tilde{C}\right) = \left(\tilde{A} \cap \tilde{B}\right) \cap \tilde{C} \tag{6.53}$$

(iii)   *Distributivity*

The distributive property describes the combined application of the union and intersection operations between three fuzzy sets $\tilde{A}$, $\tilde{B}$ and $\tilde{C}$; it follows Eqs. (6.64) and (6.65):

$$\tilde{A} \cup \left(\tilde{B} \cap \tilde{C}\right) = \left(\tilde{A} \cup \tilde{B}\right) \cap \left(\tilde{A} \cup \tilde{C}\right) \tag{6.64}$$

$$\tilde{A} \cap \left(\tilde{B} \cup \tilde{C}\right) = \left(\tilde{A} \cap \tilde{B}\right) \cup \left(\tilde{A} \cap \tilde{C}\right) \tag{6.65}$$

(iv)   *Idempotency*

$$\tilde{A} \cup \tilde{A} = \tilde{A} \tag{6.66}$$

$$\tilde{A} \cap \tilde{A} = \tilde{A} \tag{6.67}$$

(v)   *Identity*

The union and intersection of a fuzzy set $\tilde{A}$ with the null and the universe sets are defined by Eqs. (6.68)–(6.71):

$$\tilde{A} \cup \emptyset = \tilde{A} \tag{6.68}$$

$$\tilde{A} \cap X = \tilde{A} \tag{6.69}$$

$$\tilde{A} \cap \emptyset = \tilde{A} \tag{6.70}$$

$$\tilde{A} \cup X = X \tag{6.71}$$

(vi)   *Transitivity*

The transitivity property describes the relations between three fuzzy sets that mutually include them.
    If $\tilde{A} \subseteq \tilde{B}$ and $\tilde{B} \subseteq \tilde{C}$, then

$$\tilde{A} \subseteq \tilde{C} \tag{6.72}$$

(vii)   *Involution*

$$\bar{\bar{\tilde{A}}} = \tilde{A} \tag{6.73}$$

(viii)   De Morgan's principles hold for fuzzy sets as for crisp sets, as shown in Eqs. (6.74) and (6.75):

$$\overline{\tilde{A} \cap \tilde{B}} = \bar{\tilde{A}} \cup \bar{\tilde{B}} \tag{6.74}$$

$$\overline{\tilde{A} \cup \tilde{B}} = \bar{\tilde{A}} \cap \bar{\tilde{B}} \tag{6.75}$$

**Example 6.8: De Morgan's Principles Applied to Fuzzy Sets**

$$\overline{\tilde{A} \cup \tilde{B}} = \bar{\tilde{A}} \cap \bar{\tilde{B}} = \left\{ \frac{1}{1} + \frac{0}{2} + \frac{0.4}{3} + \frac{0.7}{4} \right\} \tag{6.76}$$

$$\overline{\tilde{A} \cap \tilde{B}} = \bar{\tilde{A}} \cup \bar{\tilde{B}} = \left\{ \frac{1}{1} + \frac{0.4}{2} + \frac{0.6}{3} + \frac{0.8}{4} \right\} \tag{6.77}$$

(ix)   The excluded middle laws do not hold for fuzzy sets, as described by Eqs. (6.78) and (6.79):

(a)   *Law of excluded middle: not satisfied*

$$\tilde{A} \cup \bar{\tilde{A}} \neq X \tag{6.78}$$

(b)   *Law of no contradiction: not satisfied*

$$\tilde{A} \cap \bar{\tilde{A}} \neq \emptyset \tag{6.79}$$

The excluded middle law does not hold for fuzzy sets because fuzzy sets and their complement may overlap.

## 6.4   Linguistic Variables

Consider a variable $V$ (such as the height of people, sound quality, etc.), the range of values of the variable $X$, and a finite or infinite set of fuzzy sets $T$. A linguistic variable is a triplet $(V, X, T)$. Figure 6.7 shows an example of a linguistic variable, assuming the weight of a group of people (variable), which expands from 60 to 210 kg (range) with three fuzzy sets: the low (low weight) from 70 to 130 kg, mid (mid weight) from 110 to 170 kg, and high (high weight) from 150 to 210 kg.

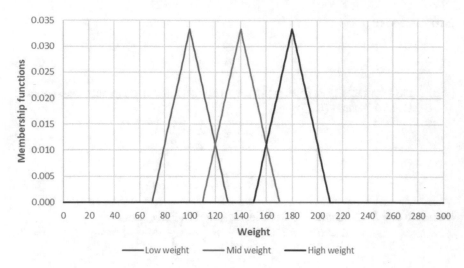

**Fig. 6.7**  Linguistic variable—people weight

The fuzzy sets are defined in the linguistic variables; they do not intend to define the linguistic variables exhaustively but define a few subsets, which are used later together with the fuzzy rules.

Fuzzy concepts use quantifiers such as cold and cool for temperature, young and old for age and short and tall for height. Concepts to be quantified with fuzzy sets are called *linguistic variables*, while the quantifiers are called *linguistic terms*.

A linguistic variable is a variable whose values are words or sentences in a natural or artificial language. A linguistic variable is used to represent any element that is either extraordinarily complex or not accurately defined; it refers to a quality, e.g., speed, temperature, project duration, etc. The range of possible values of the fuzzy variable is the universe of discourse of the linguistic variable. Fuzzy values are the fuzzy intervals used to describe the universe of discourse of the fuzzy variable.

## 6.5  Membership Functions

The membership function describes each domain element's degree of belonging (or degree of membership) in the fuzzy set. Membership functions can be of any shape, and they are determined by experts in the domain over which the sets are defined. They should satisfy two constraints: (i) their output has a lower limit of zero and an upper limit of one, a range of [0, 1], and (ii) for each value in the domain, the membership function generates only one output. Its shapes characterise the membership functions, and they have three defining characteristics, illustrated in Fig. 6.8:

**Fig. 6.8** Membership function

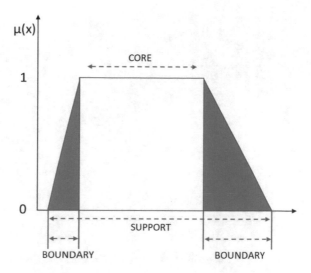

(a)  *The core*: this is the region of the membership function in which elements have full membership; if the universe is $X$, the core of a fuzzy set $\tilde{A}$ are those values $x$ of the universe for which $\mu_{\tilde{A}}(x) = 1$.

(b)  *The support*: this is the region of the membership function in which elements have membership values greater than zero; if the universe is $X$, the support of a fuzzy set $\tilde{A}$ are those values $x$ of the universe for which $\mu_{\tilde{A}}(x) > 0$. The support contains the core of the membership function.

(c)  *The boundary*: this is the region of the membership function in which elements have membership greater than zero but less than 1; if the universe is $X$, the boundary of a fuzzy set $\tilde{A}$ are those values $x$ of the universe for which $1 > \mu_{\tilde{A}}(x) > 0$. The boundary is the difference between the support and the core.

A critical point in a fuzzy set $\tilde{A}$ is the crossover value, which is the value $x$ such that, $\mu_{\tilde{A}}(x) = 0.5$.

The height $(h)$ of the membership function $\tilde{A}$ is its maximum value, $h = Max\big(\mu_{\tilde{A}}(x)\big)$.

In practical applications, the most common membership function is triangular. The Normal function is also frequently used; however, the shape is generally less important than the number of curves and their placement. Often, three to six curves are appropriate to cover the range of input values.

In some cases, the membership functions are modified or refined using "hedges" or fuzzy modifiers that are equivalent to adverbs. Hedges are fuzzy set qualifiers used to modify the fuzzy sets, e.g., "very", "quiet", "less", "likely", "extremely", "near", "close", "too", etc. create the modified fuzzy sets "very hot", "less tall", etc. Hedges create subsets on the fuzzy values, and the sets overlap, helping to reflect human thinking.

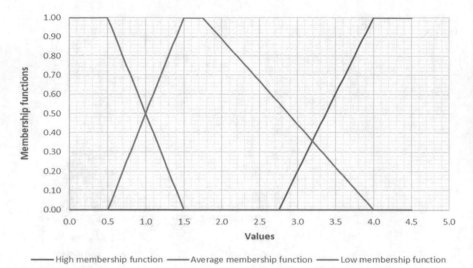

——— High membership function ——— Average membership function ——— Low membership function

**Fig. 6.9** Example of membership function

Figure 6.9 shows a set of membership functions defined using three linguistic variables: high, average, and low. The corresponding functions are shown in Appendix of this chapter.

The three membership functions determine the degree to which an element belongs to each set. For example, the number 1.20 has a degree of 0.30 in the set *low* and 0.70 in the set *average*.

There are no defined rules to build the membership function, and their shapes and parameters depend on the expert's knowledge; however, the membership function should be consistent: for example, in Fig. 6.9, the degree of belonging of value 4 in category *High* should be higher than in category *Average*, and the degree of membership of value 3 in *Average* is expected to be higher than in *High*.

## 6.6   Fuzzification Process

Fuzzification transforms crisp values to fuzzy; fuzzification accounts for the uncertainties in the crisp values through the membership functions. Uncertainties could result from imprecision in measurements, and those uncertainties are well captured with the membership function. Also, in decisions and classification problems, the uncertainty could be related to the degree to which each element belongs to the several sets; this kind of uncertainty is also well captured with the membership functions.

There are several methods to generate a membership function. However, we will skip this discussion here as it belongs to a specialised area beyond this book's scope. As a reference, neural networks, genetic algorithms, inductive reasoning, and angular

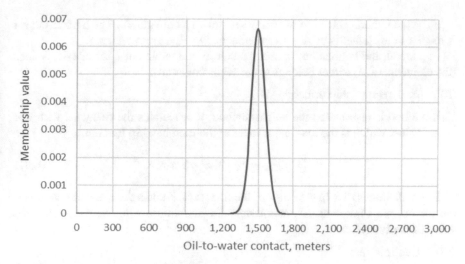

**Fig. 6.10** Fuzzy value for a measurement

fuzzy sets generate the correct membership function for specific problems. In this book, we use the intuition method, which has been the most used.

Most of the time, membership functions are generated using the expert's knowledge and understanding of the problem. Where there is uncertainty in measuring a value, the membership functions can be used to account for that uncertainty. For example, if the oil-to-water contact is measured in an electric log given an estimated value of 1,500 m and the uncertainty estimated is 10%, the membership function is described in Fig. 6.10.

## 6.7 Defuzzification Process

The defuzzification process consists of transforming fuzzy values into crisp values. For the practical application of fuzzy logic, fuzzy values should be converted to crisp values. Defuzzification transforms a collection of membership values into a single value. There are several methods of defuzzification; we discuss six of the most used methods.

(i) Lambda cuts

Lambda cuts allow a fuzzy set to be transformed into a crisp set. Consider a fuzzy set $\tilde{A}$ and define the lambda cut set $A_\lambda$, where $\lambda$ is a real number $0 \leq \lambda \leq 1$, as per Eq. (6.80):

$$A_\lambda = \left\{ x \epsilon X | \mu_{\tilde{A}}(x) \geq \lambda \right\} \tag{6.80}$$

$A_\lambda$ is the lambda cut set of the fuzzy set $\tilde{A}$ with a cut value $\lambda$. The lambda cut set is made of the elements of $\tilde{A}$ with a membership value greater than $\lambda$.

In general, the lambda cut does not result in one value but a set of crisp values (the elements of $A_\lambda$ either belongs to the set or not at all).

(ii)    Max-membership principle

This method is also called the height method; it associates the fuzzy set with one single value, which is the maximum value of the membership function:

$$\mu_{\tilde{A}}(x^*) \geq \mu_{\tilde{A}}(x) \quad for \; all \; x \in X \tag{6.81}$$

From all values of $x$ in the range of the fuzzy set, it takes the one that maximises the membership function $(x^*)$ and assigns the maximum value as the defuzzified value of the fuzzy set.

(iii)    Centroid method

The centroid is the method most frequently used for defuzzification. The centroid method calculates the centre of the area using the membership function's values as the weighting factor. This method aggregates the results for all values represented by the membership function; the defuzzified value is calculated using Eq. (6.82):

$$x^* = \frac{\int \mu_{\tilde{A}}(x) \cdot x \cdot dx}{\int \mu_{\tilde{A}}(x) \cdot dx} \tag{6.82}$$

This method generates the average value of the variable $x$ using the membership function as the weight factor. It is also called the centre of gravity, the centre of the area or the centre of mass method.

(iv)    Weighted average method

This method works when the resulting fuzzy set is the aggregation of two or more symmetrical fuzzy sets. In this case, the weighted average method is the average of the values with higher membership functions per each fuzzy set combined to produce the final fuzzy set weighted with their corresponding membership function values. It is given by Eq. (6.83):

$$x^* = \frac{\sum \mu_{\tilde{A}}(\overline{x}) \cdot \overline{x}}{\sum \mu_{\tilde{A}}(\overline{x})} \tag{6.83}$$

(v)    Mean-max membership

This method applies well when the maximum value of the membership function is not at one point but within an interval; in this case, if $a$ and $b$ are the minima and maximum values in the range where the membership has its maximum value, then

$$x^* = \frac{a + b}{2} \tag{6.84}$$

## (vi) Mean of maximum

The mean of maxima considers only the part of the fuzzy output set with a maximal membership value. The value obtained with this method is the average of the interval where the function gets its maximum value. This method cannot be applied directly in the multimodal membership function; when more than one interval has the maximum value, the mean-max membership is applied to each interval. For a unimodal membership function, the defuzzified value for this method is computed following Eqs. (6.85), (6.86) and (6.87),

$$x_{min} = inf\{x \in X | \mu_{\tilde{A}}(x) = max\{\mu_{\tilde{A}}(x)\}\} \tag{6.85}$$

$$x_{max} = sup\{x \in X | \mu_{\tilde{A}}(x) = max\{\mu_{\tilde{A}}(x)\}\} \tag{6.86}$$

$$x^* = \frac{x_{min} + x_{max}}{2} \tag{6.87}$$

## (vii) Centre of sum

The centre of sum method adds the individual output fuzzy sets. The centre of sum method is similar to the centre of area method. Still, it does not consider the aggregation for the complete fuzzy set output but adds the resulting of each fuzzy set evaluation individually. This method has the disadvantage of double counting the values where there is overlapping between the fuzzy sets, which happen because of the rules' fuzziness. The great advantage of this method is easy to implement the algorithms and the fast computing of the results. The centre of the sum is calculated using Eq. (6.88), and the defuzzified value $x^*$ is given as

$$x^* = \frac{\int x \cdot \sum_{k=1}^{n} \mu_{\tilde{A}_k}(x) \cdot dx}{\int \sum_{k=1}^{n} \mu_{\tilde{A}_k}(x) \cdot dx} \tag{6.88}$$

In Eq. (6.86), the range of the fuzzy sets is divided into n intervals, and the membership is evaluated in each interval. The summation can be replaced by an integral to improve the accuracy of the estimation; in this case, the method involves the evaluation of a double integral.

## 6.8 Fuzzy Rules

Fuzzy rules are built to implement inference systems based on linguistic variables for the antecedents and the consequents. The antecedent part of the rule states an inference that, if satisfied, the rule is true; the consequent part of the rule is the conclusion if the antecedent is true.

Fuzzy rules were first formulated by Zadeh (1975) for analysing complex systems by capturing human knowledge. Fuzzy rules are made of statements describing how the inputs are combined, using fuzzy sets, to produce a fuzzy output; this output is used for making decisions or control systems. Rules are made of IF …..THEN pairs of statements. Based on the set of rules, the microcontroller makes decisions for what action to take.

Rule creation

The rules are created using three methods:

(i)  *Assignment statements*: these are rules created using the assignment operator " = "; thus, a value (either numeric or linguistic) is assigned to a variable. Examples of this rule are:

- The weather is cold.
- John is tall.
- $B = 3$.

(ii)  *Conditional statements*: a conditional statement is a statement that is considered valid if another statement is satisfied. Examples of this rule are:

- If John is 1.80 m, then he is tall.
- If the car speed is 100 km/h, then the speed must be reduced.
- If the temperature is higher than 30 °C, then it is hot.

These statements are conditional because if a condition is fulfilled, then a restriction applies.

(iii)  *Unconditional statements*: like the conditional statement but without the condition. They take the form of an order. Examples of unconditional statements are:

- Increase in speed.
- Open the door.
- Turn on the light.

Both conditional and unconditional rules set restrictions on the consequence of the rule-based system.

In the context of set theory, a fuzzy rule is a conditional statement on the linguistic variables $x$ and $y$ over the linguistic values $A$ and $B$ as follows:

- IF is THEN is

where and are sets determined on the universe of discourses and, respectively.

IF….THEN rules are not exclusive to fuzzy sets; indeed, they have been applied extensively on classical sets; however, those are different from fuzzy rules. IF….THEN rules for classical sets have the following form:

*Rules 1:*

- IF the room temperature is (equal to/lower than) 25 °C

- THEN the air conditioning is turned on
- IF the room temperature is higher than 25 °C
- THEN the air conditioning is turned off

The IF statement is the antecedent, and the THEN statement is the consequence of the rule.

The crisp variable (antecedent) "room temperature" measures the temperature in the room and, depending on whether it is (equal to/lower than) or higher than 25 °C, the air conditioning is turned on or turned off, respectively; that is, the values (consequence) are crisp.

These kinds of rules are Boolean, e.g., the condition stated in the "antecedent" (the IF statement") either entirely belongs or does not belong at all to the "consequence" (the THEN statement").

These rules can be modified as below:

*Rules 2:*

- IF the room temperature is (equal to/lower than) 25 °C
- THEN the weather is pleasant
- IF the room temperature is higher than 25 °C
- THEN the weather is unpleasant

Comparing rules 1 with 2, we keep the antecedent part the same but modify the consequence part (the weather) from crisp to fuzzy values.

These rules can be further modified to have fuzzy variables and fuzzy values:

*Rules 3:*

- IF the room temperature is low
- THEN the weather is pleasant
- IF the room temperature is high
- THEN the weather is unpleasant

In the example using rule 3, both the antecedents and the consequence parts of the rule are fuzzy: temperature measured as "low" and "high" has a fuzzy meaning, and the weather being "pleasant" or "unpleasant" is also a fuzzy term.

Conditional, unconditional, and restrictions are tied using linguistic connectors like "AND", "OR", "ELSE". Typically, "AND" uses all the antecedents' minimum value, while "OR" uses the maximum value. An example of a canonical form for the rules is:

- IF conditions "AND" then restriction

Fuzzy rules can have more than one antecedent:

*Rules 4:*

- IF the room temperature is low
- AND the humidity is medium
- THEN the weather is comfortable

Fuzzy rules can have more than one consequence.

- IF the room temperature is low
- THEN the weather is nice
- AND people feel hungry

A fuzzy expert system typically has several rules; each rule's output is also a fuzzy set that can be transformed into a number, as will be shown later.

The main difference between the classical and fuzzy rules is that, in the classical rules, when the antecedent is true, the consequence is true. The antecedent triggers the result. The antecedent partially fires several implications in fuzzy rules to some degree, depending on the membership functions.

Rule aggregation

When there is more than one rule, each rule's consequent should be aggregated to produce an overall conclusion from the system. Aggregation can be done in two ways:

(i)   *Conjunctive method*: by using the connector "AND", the several outputs are aggregated by intersecting the rule consequents:

$$x = x^1 \, AND \, x^2 \, AND \dots AND \, x^n \tag{6.89}$$

Equation (6.87) can be written using a membership function, as per Eq. (6.90):

$$\mu_x(x) = \min(\mu_{x^1}(x), \mu_{x^2}(x), \dots, \mu_{x^n}(x)) \tag{6.90}$$

(ii)  *Disjunctive method*: the rules outputs are aggregated by the connector "OR". The overall result is obtained by the fuzzy union of all the consequents:

$$x = x^1 \, OR \, x^1 \, OR \dots OR \, x^n \tag{6.91}$$

Using membership functions, the disjunctive method is

$$\mu_x(x) = \max(\mu_{x^1}(x), \mu_{x^2}(x), \dots, \mu_{x^n}(x)) \tag{6.92}$$

## 6.9  Fuzzy Inference Systems

A fuzzy inference system (FIS) is the process of mapping input variables into output variables utilising fuzzy rules. A FIS is formulated based on fuzzy set theory, fuzzy IF…THEN rules, connectors such as "AND" and "OR", combining the rules and fuzzy reasoning. The output of the FIS is used to make a decision or control a system.

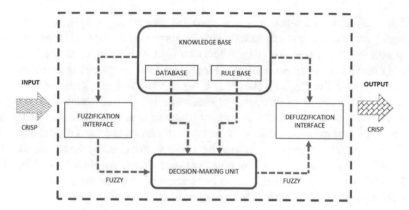

**Fig. 6.11**  Fuzzy inference system

A FIS can use fuzzy or crisp inputs, and the outputs are fuzzy sets; however, the outcomes can be defuzzified to get a crisp number representing the fuzzy output.

Figure 6.11 shows a schematic of a FIS with its constituent parts.

As described in Fig. 6.11, a FIS has five components described below.

i.   *Fuzzification interface*: transforms crisp inputs into degrees (values) within a fuzzy set (linguistic variable).
ii.  *Rule base*: this is the IF…THEN rules that define the FIS.
iii. *Database*: consists of the membership functions for the different fuzzy sets.
iv.  *Decision-making unit*: this part of the FIS makes the inference operations on the rules.
v.   *Defuzzification interface*: transforms the fuzzy results of the inference into crisp outputs.

The inputs values are applied over the input membership functions, and the rules are applied. The rules (rules strength) results are mapped over the output membership functions to get the rules' outcome. These rules results are aggregated, and the combined result is defuzzified, resulting in a crisp output value that is the final evaluation of the system.

Fuzzy inference methods

The most frequently used fuzzy inference methods are *Mamdani´s fuzzy inference method* and the *Sugeno or Takagi–Sugeno-Kang method*. The Mamdani method was introduced by Mamdani (1974) and Mamdani and Assilian (1975), and it uses fuzzy sets as rule consequents which, if necessary, can be defuzzified to obtain a crisp number. The Takagi–Sugeno-Kang or TS method was introduced by Takagi and Sugeno (1985), and it uses linear functions of input variables as rule consequents. The existing applications of fuzzy systems are developed using Mamdani fuzzy systems; for this reason, in this book, we will focus on the Mamdani method only.

Mamdani fuzzy inference was first used for developing a system to control a steam engine, and boiler combination using 24 linguistic control rules defined based on operators' experience. The model has two inputs: heat input to the boiler and throttle opening at the engine cylinder's input, and two outputs: the steam pressure in the boiler and the engine's speed. Six fuzzy variables and seven subsets were defined in the universe of discourse of the variables. A fixed controller was used for comparison proposes; it was run many times to tune the controller for the best performance. Comparing the fuzzy controller with a standard fixed digital one found that the fuzzy control was better than the best control obtained by the fixed controller. This outstanding paper applied the concepts that Zadeh developed (Zadeh, 1965, 1973, 1974) about fuzzy algorithms for complex systems control and decisions.

In the Mamdani method, after the aggregation of the rules' outcomes, the resulting product is a fuzzy set; if required to have a crisp result, any of the defuzzified methods discussed in Sect. 6.7 can be used.

Alternatively, it is sometimes possible to use a single spike as the output membership function instead of a typical fuzzy function; this may simplify the calculation associated with the defuzzification; this type of output function is called a *singleton*. However, with the current extensive use of computers and sophisticated algorithms, complex functions' defuzzification is not a real challenge.

The Mamdani fuzzy inference method has six steps:

i.    *Determine the set of fuzzy rules* associated with the problem concerned; rules are made of combinations of fuzzy sets (membership functions), each corresponding to a linguistic variable.
ii.   *Fuzzify the input variables* using the membership functions; this step is carried out by applying the membership function to the crisp input values.
iii.  *Combine the fuzzified inputs* using the fuzzy connectors (AND/OR) following the fuzzy rules to get the *rule strength*.
iv.   Find the *rule's consequences* by combining the *rule strength* with the *output membership function (consequence)*; this is done by clipping the output membership function at the rule strength value. The rule consequence is a fuzzy set.
v.    *Combine the consequences* (fuzzy sets) to get a *fuzzy output* (or fuzzy consequence); this step is called *aggregation*.
vi.   Perform a *fuzzy output defuzzification* to get a *crisp output* (if a crisp result is required).

The *set of fuzzy rules* is a combination of linguistic statements built to enable the FIS to decide on the classification or control of outputs. Fuzzy rules have the form:

IF (input 1 has membership in a given fuzzy set) and/or (input 2 has membership in a given fuzzy set) THEN (the output has membership in a given fuzzy set).

IF the temperature is cold AND the humidity is low, THEN the weather is good.

*Fuzzification*, step (ii) of the Mamdani fuzzy inference method is the process of mapping the set of crisp input values (a crisp value resulting from a measurement, sensor, etc.) into the set of real numbers [0, 1] by using the input membership functions. As explained in Sect. 6.6, the membership values result from intersecting the value with the membership function; the membership function measures the linguistic variable's fuzziness to describe reality.

To *combine the fuzzified inputs* in a rule, the connectors AND/OR are used; these connectors represent the operations' maximum or minimum. The strength of the rule results from applying the connectors over the fuzzified inputs.

The *rule's consequence* is obtained by clipping the consequent (output) membership function at the rule strength value.

*Combining the outputs or aggregation* to get an output distribution is usually done using the connector "OR" and sometimes "AND", as discussed in Sect. 6.8. In this step, each rule's fuzzy consequence is aggregated to generate the system's fuzzy output. This step could be the final one if the system result can be a fuzzy set. However, if the result is required to be a crisp number, then the fuzzy product should be defuzzified as described in the next step.

Defuzzification of the output set is necessary when the system's result needs to be a crisp value. Several defuzzification techniques were described in Sect. 6.7; the most used are Centroid and Mean of maximum.

**Example 6.9: Development of a FIS for Decision Making Using MATLAB®** In this example, we build a FIS to classify whether the weather is comfortable or not based on two parameters, temperature and humidity. This system can be used for decision-making. As the control system, the decision-making system FIS combines the temperature and humidity to decide, within the fuzzy criteria, if a given set of temperature and humidity corresponds to a certain category of the comfort level. As a control system, it can set a device such as air-conditioning to adjust its functioning for people's comfort.

The author decided to use triangular membership functions to capture the fuzziness in the input and output of the data because they are defined, univocally, with three numbers—the minimum, maximum and medium of each variable, which reflects the author's ranges for their variability.

The shape of the membership functions and the values used in this example is a given selection that can be modified according to personal criteria.

Figure 6.12 depicts the FIS of this example, where the Mamdani method is used.

As mentioned before, the two input variables, temperature and humidity, use triangular functions; the temperature is described by three fuzzy functions LowTemp (Low Temperature), MidTemp (Medium Temperature) and HigTemp (High Temperature). Similarly, humidity is defined by the membership functions LowHum (low humidity), MidHum (Medium humidity) and HigHum (High Humidity). Each function's parameters are selected according to the author's criteria for the variability of the linguistic variables associated with the input variables. Figure 6.13 describes the two input variables and the six membership functions.

The output functions represent the person's degree of satisfaction in the experiment when exposed to the complete range of temperature and humidity described in the input membership function. Figure 6.14 refers to the output functions as Pleasant

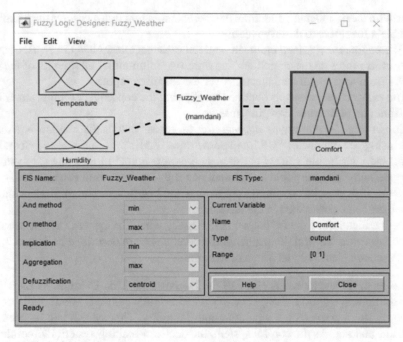

**Fig. 6.12** Fuzzy inference system for weather decision

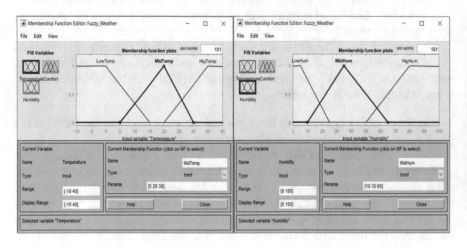

**Fig. 6.13** Fuzzy input membership functions for the inference system for weather decision

(satisfactory, good), Regular (intermediate) and Unpleasant (not satisfactory, bad); the author defines the range of values associated with each membership function according to his beliefs. Figure 6.14 depicts the output membership functions.

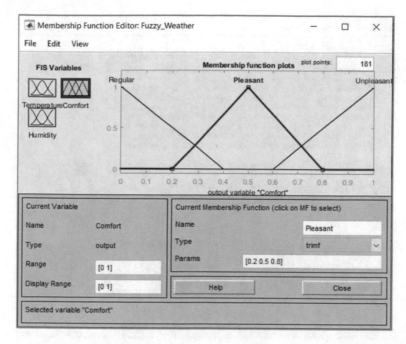

**Fig. 6.14** Fuzzy output membership functions for the inference system for weather decision

The FIS rules contain the combinations of the input variables that produce a given output. In this example, nine rules define how the two input membership functions combine to make the nine output rules, as shown in Fig. 6.15.

Figure 6.16 shows three evaluations of the FIS for given selections of temperature and humidity. In the upper plot, the temperature is 15 °C, and the humidity is 50%, resulting in an output value of comfort of 0.5. In the left-bottom plot, the temperature is 5 °C, and the humidity is 20%, producing a level of comfort of 0.431. In the right-bottom plot, the temperature is 30 °C, and the humidity is 5%, creating a comfort level of 0.858.

This example shows the implementation of a FIS using two input variables in a Mamdani inference system to control an air-conditioning system or make decisions regarding comfort level. The software used is MATLAB® (@MatWorks), but other software can also be used for the same purpose.

## 6.10   Fuzzy Applications

Many appliances, from a dishwasher to a coffee machine, use fuzzy logic in their day-to-day functions. However, fuzzy logic has also been used for more than 30 years in highly critical machinery and systems that operate in a complex and changing environment.

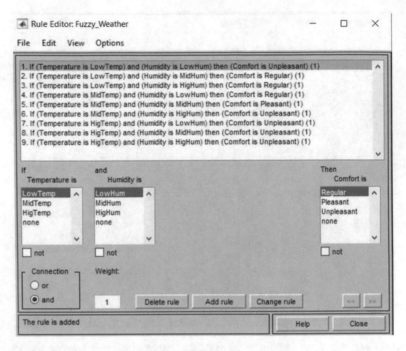

**Fig. 6.15**  Fuzzy rules for the inference system for weather decision

In 1978 the Danish company F. L. Smidth achieved the control of a cement kiln using a fuzzy system which was the first genuine industrial application of fuzzy logic.

In 1985 two engineers from Hitachi, S. Yasunobu and S. Miyamoto, developed a control system that demonstrated the feasibility of using fuzzy logic to control the Sendai Subway Namboku Line railway, Japan. This concept was later proved in practice when the railway started operations in 1987. The fuzzy system accelerated, controlled the braking system and camcorder focusing system, and stopped the train. This fuzzy system allows a smooth start and stop compared with other trains and is 10% more energy-efficient than human-controlled trains; passengers hardly notice when the train is changing its velocity.

In 1988, Japan's government established the Laboratory for International Fuzzy Engineering (LIFE), dedicated to developing a fuzzy system for industrial and consumer applications; 48 companies have been involved in this venture.

Hitachi also incorporates fuzzy logic in one of its washing machines to control the load-weight, dirt sensors, fabric type and select the optimum wash cycle for the best use of power, water and detergent.

Canon developed a fuzzy control system to optimise the camera lens position based on the image's clarity (as measured by the charge-coupled device—CCD) and the change of lens movement during focusing. The autofocusing function based on fuzzy logic has been massively implemented in many camera models by Canon and Minolta.

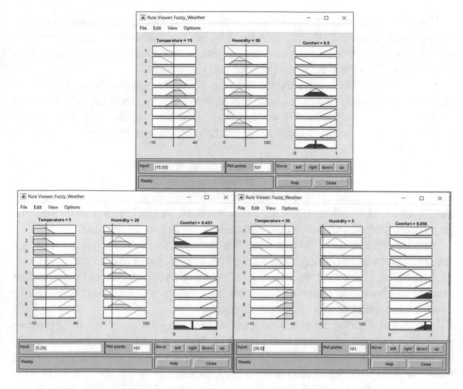

**Fig. 6.16** Fuzzy crisp evaluation of the inference system for weather decision

Mitsubishi developed a fuzzy control system for industrial air conditioning systems using 25 heating rules and 25 cooling rules; the rules' trigger is a temperature sensor. The outputs control an inverter, a compressor valve, and a fan motor. This fuzzy controller heats and cools five times faster, reduces energy consumption by 24% and increases the temperature stability twofold.

In the literature, there are many references to fuzzy logic implementation. Ivanova and Zlatanov (2019) discuss a successful application of fuzzy logic in online test evaluation in English as a foreign language, specifically for students with qualifications on the borderline between two grades, without impacting the overall group score. Darwich et al. (2018) applied fuzzy logic to detect the moving foreground by subtracting the image background using several fuzzy classifiers to estimate the pixel class for each frame.

Ariani and Endra (2013) implement a FIS for deciding the optimum study programme selection. The successful application of this approach secures student motivation during the studies, avoiding student failure. The input variables consider different forms of a scoring system, and the output is the students' interest in either the Department of Informatics Engineering or the Department of Information System.

Greeda et al. (2018) discuss the use of fuzzy logic and fuzzy systems in several areas of medicine, such as during anaesthetics; also, they consider the implementation of fuzzy expert systems in the medical diagnosis of a variety of illnesses such as tuberculosis, infantile cancer, lung cancer, and diabetes.

Vilela et al. (2018) develop the formulation of the methodology of VOI, incorporating the uncertainty due to the fuzziness in the data. This methodology is further applied to a case study related to the decision alternatives of either moving forward with the development of an appraisal oil field, relinquishing it, or acquiring additional data before deciding.

Applications of fuzzy logic and fuzzy controllers have been in many different domains. Pappis and Mamdani (1977) propose implementing a fuzzy logic controller in a single intersection of two one-way streets. The implemented model showed that the performance criterion, measured as the vehicles' delay, is much better with the fuzzy controller than with the conventional vehicle-actuated controller.

Van De Ville et al. (2003) present an application of fuzzy logic for image processing, particularly image filtering by noise reduction of images corrupted with additive noise. In this case, the filtering is developed in two stages, and the complete filtering can be applied iteratively to reduce the heavy noise. Experimental results show the feasibility of this filtering technique.

Vilela et al. (2019a) discuss the impact of including the fuzzy nature of the data in the VOI assessment. A case study is used to prove that by adding the uncertainty related to the fuzziness, the value of the data is reduced. In the case study discussed, while the VOI for crisp data is favourable compared with the no data acquisition scenario, when the fuzziness of the data is incorporated, the VOI for fuzzy data is negative.

Maryam and Khooban (2013) design a Mamdani-type fuzzy logic controller for trajectory tracking of wheeled mobile robots considering parametric and nonparametric uncertainties. The model was implemented using the MATLAB®/Simulink environment. The simulation results show that the system performance is acceptable.

The risk assessment associated with pipelines is a critical task due to the significant impact of pipeline services disruption. These pipelines can transport liquid and gas; hydrocarbon products transportation is one of the most recurrent uses of pipelines. Jamshidi et al. (2013) use fuzzy logic for the pipeline's risk assessment considering the uncertainty due to a lack of information and uncertainty in modelling. The model is built using the fuzzy logic toolbox of MATLAB and the Mamdani method; a case study is discussed. The results demonstrate the value associated with using FIS for the risk assessment.

One of the significant hazards in underground coal mines is mine fire; continuous monitoring of risk factors such as temperature, relative humidity, the concentration of different gases, etc., is crucial. Using a wireless sensor network is extremely important for securing coal mine integrity and worker safety. Muduli et al. (2018) propose a wireless sensor network using fuzzy logic to enhance the decision-making process's reliability during the mine's operation. The Mamdani inference system was implemented using MATLAB. The results show that the proposed fuzzy system is more reliable for fire hazards than the offline monitoring system typically used in coal mines.

Vilela et al. (2019b) develop a FIS for assessing the value of information in decision problems. A case study related to the possible development of an oil field with high uncertainty in well productivity shows the benefit of this approach. Two outcome parameters (net present value and discounted profit to investment ratio) are used to assess the project. Using the standard methodology, the conclusion reached is ambiguous because both parameters conduct to contradictory results. Then, by implementing the FIS, we can manage the linguistic terms that decision-makers use to assess the project. It also allows to aggregate, successfully, both fuzzy outcome parameters into one crisp parameter that provides a clear decision outcome.

## 6.11 Summary

Chapter six begins with a historical review of the origin of fuzzy logic and the reasons that justify its introduction as a complement to classical logic. We review the crucial ideas that took shape in the work of Zadeh. Further, we additionally review the classical set theory and then introduce the fuzzy set theory as a complement to the former. The main distinction between classical and fuzzy set theories is captured by the excluded middle laws that do not hold in fuzzy theory. A review of the linguistic variable concept is given, which is critical in understanding fuzzy sets. Another way to contrast these theories (classical and fuzzy) is through the characteristic/membership functions; a discussion of membership functions, properties, and characteristics is also included. The meaning, importance, and methods of fuzzification and defuzzification are discussed. We introduce how fuzzy rules are made and why they are essential, including a few examples. We follow the chapter with the important subject of the fuzzy inference system, which is a very popular method for implementing fuzzy logic in practical applications. We also include a detailed description of the Mamdani inference method and discuss a complete example of a fuzzy inference system in MATLAB software for a real case example. The chapter concludes with a review of applications of fuzzy logic in several domains. The main reason for including fuzzy logic in this book is that it can account for the uncertainty associated with the fuzziness in the data; this uncertainty can greatly impact the value of the data and should be included in the value of information workflow, as demonstrated in Chap. 7.

## Appendix

The membership functions corresponding to Fig. 6.9 are described in Eqs. (6.93)–(6.95). The three functions—High, Average, and Low—describe the degree of belonging of the linguistic variable's values.

$$\mu_{High}(x) = \begin{cases} 0 & x \le 2.75 \\ \frac{(x-2.75)}{1.25} & 2.75 < x \le 4 \\ 1 & x > 4 \end{cases} \tag{6.93}$$

$$\mu_{Average}(x) = \begin{cases} 0 & x \le 0.5 \\ \frac{(x-0.5)}{1} & 0.5 < x \le 1.5 \\ 1 & x > 1.5 \\ \frac{(4-x)}{2.25} & 1.75 < x \le 4 \\ 0 & x > 4 \end{cases} \tag{6.94}$$

$$\mu_{Low}(x) = \begin{cases} 0 & x \le 0.5 \\ \frac{(1.5-x)}{1} & 0.5 < x \le 1.5 \\ 1 & x > 1.5 \end{cases} \tag{6.95}$$

# References

Ariani, F., & Endra, R. (2013). Implementation of fuzzy inference system with Tsukamoto method for study programme selection. In *Proceeding of the 2nd International Conference on Engineering and Technology Development (ICETD 2013)* (pp. 189–200). Faculty of Engineering and Faculty of Computer Science, Bandar Lampung University.

Darwich, A., Hébert, P. A., Bigand, A., & Mohanna, Y. (2018). Background subtraction based on a new fuzzy mixture of Gaussians for moving object detection. *Journal of Imaging, 4*, 92. https://doi.org/10.3390/jimaging4070092.

De Ville, D., Nachtegael, M., Van der Weken, D., Kerre, E., Philips, W., & Lemahieu, I. (2003, August). Noise reduction by fuzzy image filtering. *IEEE Transactions on Fuzzy Systems, 11*(4), 429–436.

Greeda, J., Mageswari, A., & Nithya, R. (2018). A study on fuzzy logic and its applications in medicine. *International Journal of Pure and Applied Mathematics, 119*(16), 1515–1525.

Ivanova, V., & Zlatanov, B. (2019). Application of fuzzy logic in online test evaluation in English as a foreign language at university level. *AIP Conference Proceedings, 2172*, 040009. https://doi.org/10.1063/1.5133519.

Jamshidi, A., Yazdani-Chamzini, A., Yakhchali, S., & Khaleghi, S. (2013). Developing a new fuzzy inference system for pipeline risk assessment. *Journal of loss prevention in the process industries, 26*, 197–208.

Mamdani, E. (1974). Application of fuzzy algorithms for control of simple dynamic plant. *Proceedings of the Institution of Electrical Engineers, 121*(12), 1585–1588. https://doi.org/10.1049/piee.1974.0328

Mamdani, E., & Assilian, S. (1975). An experiment in linguistic synthesis with a fuzzy logic controller. *Int. J. Man-Machine Studies, 7*, 1–13.

Maryam, D., & Khooban, M. (2013). Design of optimal Mamdani-type fuzzy controller for nonholonomic wheeled mobile robots. *Journal of King Saud University-Engineering Science, 27*(1), 92–100.

Muduli, L., Jana, P., & Mishra, D. (2018). Wireless sensor network-based fire monitoring in underground coal mines: A fuzzy logic approach. *Process Safety and Environmental Protection, 113*, 435–447.

Pappis, C., & Mamdani, E. (1977). A fuzzy logic controller for a traffic junction. *IEEE Transactions on Systems, Man, and Cybernetics, SMC-7*(10), 707–717.

Takagi, T., & Sugeno, M. (1985). Fuzzy identification of systems and its applications to modeling and control. *IEEE Transactions on Systems, Man, and Cybernetics, SMC-15*(1), January/February.

Vilela, M., Oluyemi, G., & Petrovski, A. (2018). Fuzzy Data analysis methodology for the assessment of value of information in the oil and gas industry. In *Proceeding of the 2018 IEEE International Conference on Fuzzy Systems (FUZZ-IEEE)* (pp. 1540–1546).

Vilela, M., Oluyemi, G., & Petrovski, A. (2019a). Fuzzy logic applied to value of information assessment in oil and gas projects. *Springer, Petroleum Science, 16*(5), 1208–1220.

Vilela, M., Oluyemi, G., & Petrovski, A. (2019b). A fuzzy inference system applied to value of information assessment for oil and gas industry. *Decision Making: Applications in Management and Engineering, 2*(2), 1–18.

Zadeh, L. (1965). Fuzzy sets. *Information and Control, 8*, 338–353. https://doi.org/10.1016/S0019-9958(65)90241-X

Zadeh, L. (1973). Outline of a new approach to the analysis of complex system and decision process. *IEEE Transaction on Systems, Man, and Cybernetics, SMC-3*(1), 28–44.

Zadeh, L. (1974). Application of fuzzy algorithms for control of simple dynamic plant. *Proceedings of the Institution of Electrical Engineers—Control & Science, 121*(12), 1585–1588.

Zadeh, L. (1975). The Concept of a Linguistic Variable and its application to Approximate Reasoning–I. Information Sciences *8*, 199–249.

# Chapter 7
# Uncertainty, Data Acquisition and Value of Information Assessment

**Objective**

In this chapter, we discuss the methodology of the value of information which is one of the central topics of this book (together with the value of flexibility, Chap. 8). The theoretical and mathematical formalism is presented, and several examples are discussed. The logic behind the value of information, for increasing the value of a project through improved chances of success, is explained. The values of perfect, imperfect and fuzzy data are discussed.

## 7.1   Introduction to the Value of Information

As mentioned in Chap. 1 of this book, uncertainty is a characteristic of a variable not being known with certainty. This uncertainty at present will, of course, brings uncertainty to the variable's value in the future.

Most decision problems involve uncertainties, and, in many cases, the analysis of uncertainties is carried out not to reduce them but to understand them better and assess their impact on the decisions. Therefore, in a decision analysis process, the following steps are required.

(i)    Uncertainties should be identified and documented; we need to recognise which have more influence on the decision.
(ii)   Uncertainties should be characterised by assigning probabilities.
(iii)  The consequences of using a selected probability distribution for managing alternatives should be estimated; if the results are insensitive to several combinations of probability distributions, either the uncertainty is irrelevant, or the valuation needs to be re-assessed.
(iv)   The distribution of outcomes has some options that are riskier than others.

M. J. Vilela and G. F. Oluyemi, *Value of Information and Flexibility*,
Petroleum Engineering, https://doi.org/10.1007/978-3-030-86989-2_7

(v)     The level of agreement or disagreement between experts when assigning probabilities should be analysed.
(vi)    Options to reduce uncertainties are essential to be considered in the analysis (value of information).

Decisions are straightforward when faced with decisions with no uncertainties (that situation is extremely unlikely to occur in the real world). These decision problems consist of a set of alternatives, input variables and output variables. The system or problem at hand defines the input variables. The output variables measure what is relevant to the decision-maker, such as production, money or other forms of benefit. If our problem does not have uncertainty, the alternative with the higher value is the best decision.

Making decisions becomes more convoluted when there are uncertainties; when some or all of the inputs have uncertainties, those uncertainties translate to uncertainties in the outputs. e.g., in the development of an oil field, uncertainties in the initial rate of the wells or the reservoir permeability values translate to uncertainties in the oil production forecast of the wells and, consequently, in the complete field production.

Acquiring additional data can, in some cases, contribute to a better understanding of the uncertainties and eventually reduce them. However, the justification for data acquisition does not rely on reducing uncertainties but on improving what matters to the decision-maker, the value of the project.

In this chapter, we will use the concept of a *"project"* to refer to an endeavour that requires an initial investment (CAPEX) to generate benefits in a defined time frame; additional cost (OPEX) is required during the defined period to continue the operation.

## 7.2   The Meaning of the Value of Information

*Value of information* (VOI) is the methodology to assess, before making a decision, whether acquiring new information will bring value to the project. The value of a project is calculated by using the value of the future states of nature, the likelihood of those states before the possible data acquisition, the reliability of the data to predict, accurately, the future states of nature, and the cost associated with the acquisition of data whether that cost corresponds to money that should be invested to acquire the data or to delay the project execution due to the time required to obtain the data.

Very frequently, people think of data acquisition as a means of reducing uncertainty, and that is partially right; data acquisition provides information on the system or process under study and, as such, gives the analyst a better understanding of the uncertainties of the parameters or variables involved in the problem under investigation. However, beyond the academic world, what drives the decisions about data acquisition is the financial value (monetary) or, in a broader sense, the utility value of that data.

As mentioned by Bratvold et al. (2007),

(i)    VOI should assess the benefits of gathering new information before making a decision.
(ii)   When VOI is made, there should be the possibility of changing the decision that would be made before acquiring the data.

Statement (i) refers to the property that VOI's main concern is not reducing the uncertainty but the additional value that the new data proposed to acquire provide to the project.

Statement (ii) highlights that if a decision has already been taken based on other considerations (strategy, feelings, etc.), there is no value in acquiring new data. If new data are acquired, then the decision about the project's future should be made exclusively based on the VOI assessment.

Data acquisition takes several forms. For example, in a subsurface problem of the oil and gas industry, data acquisition can be the drilling of a well and measuring oil saturation or, in general, measuring or obtaining reservoir properties at the well location. However, it could also be an analysis (study) of analogous fields or study modelling; in other domains, data acquisition can take the form of a survey, mathematical analysis or research.

## 7.3    The Formalism of the Value of Information

Before discussing the VOI methodology, we will review the mathematics of making decisions in the framework of the Bayesian approach; later, we will include, in this formalism, the possibility to acquire new information.

As mentioned in Chap. 2 of this book, we decide on events that will happen in the near or far future, but the decision is made today, and several alternatives should be available.

In most real-world problems, the current values of the input variables have uncertainties. Consequently, the future values of the output variables also carry uncertainties, which makes it challenging to make a decision.

In the classical Bayesian decision theory framework, we select a few discretised "*states of nature*" contained in the set of states $S = \{s_1, s_2, \ldots, s_n\}$. To each state of nature, we assign a probability such that,

$$\sum_{1=1}^{n} p(s_i) = 1 \tag{7.1}$$

where $p(s_i)$ is the probability of occurrence of the state $s_i$.

These probabilities shown in Eq. 7.1 are called *prior probabilities*. They quantify the current knowledge of the likelihood for each *state of nature* to occur; they are *prior* to any further information that can be acquired.

Let assume that the decision-maker has $m$ alternatives to choose, $A = \{a_1, a_2, \ldots, a_m\}$; for any given alternative $a_j$, we assign the utility value $u_{ji}$ if the future *state of nature* turns out to be the state $s_i$.

The terms $u_{ji}$ represent the utility value (or value for risk-free evaluation) for the paired alternative $a_j$ and state $s_i$.

The expected utility value (EUV) of the alternative $a_j$ is defined as

$$E(a_j) = \sum_{i=1}^{n} u_{ji} p(s_i) \tag{7.2}$$

The maximum EUV among all the alternatives is the typical decision criterion used for making decisions:

$$E(a^*) = \frac{max}{j} E(a_j) \tag{7.3}$$

where $a^*$ is the alternative that maximises the expected value.

**Example 7.1: Decision Problem** Let us consider the case where the decision-maker is confronted with two alternatives, *alternative 1* and *alternative 2,* whose states, prior probabilities and utilities are shown in Table 7.1.

Following Eq. (7.2), the EUV of alternative 1 is 62.5, and the EUV of alternative 2 is 55.5; those calculations mean that the best alternative, Eq. (7.3), the one that maximises the EUV, is alternative 1.

In this example, each alternative represents investment alternatives for the decision-maker; these may refer to different projects or the same project under different scenarios.

This assessment can be represented using a decision tree, as shown in Fig. 7.1. Conventionally, decision nodes are represented by squares, circles represent probability nodes, and diamonds represent terminal nodes.

**Table 7.1** Example 7.1: description of the alternatives

| Alternative | State | Prior probability | Utility/Value |
|---|---|---|---|
| Alternative 1 | $s_1$ | 0.20 | 160 |
| | $s_2$ | 0.35 | 100 |
| | $s_3$ | 0.45 | −10 |
| Alternative 2 | $s_1$ | 0.15 | 250 |
| | $s_2$ | 0.25 | 120 |
| | $s_3$ | 0.60 | − 20 |

**Fig. 7.1** Decision tree for a
decision problem

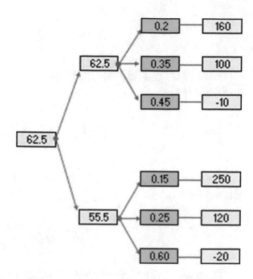

## 7.4   Value of Information

There are many situations where, apart from the decision alternatives that the
decision-maker has, there is also the possibility to acquire additional data to
characterise the alternatives under consideration better.

Suppose that new information can be acquired about the variables affecting the
utilities or values and let us call this dataset $X = \{x_1, x_2, \ldots, x_k\}$ where $k$ pieces
of data can be acquired. Using Bayes' theorem, already discussed in Chap. 4 of this
book, the new information can update the prior probabilities. The process of updating
the prior probabilities has two steps:

(i)   $T$ he probability that the new information $x_k$ confirms that the true *state of
      nature* is $s_i$ is $p(x_k|s_i)$; these conditional probabilities are known as likelihood
      values or reliability probabilities; they refer to the probabilities we assign to
      the data to predict or estimate the states of nature accurately.
(ii)  Given that the new information $x_k$ is true, the probability that the true *state
      of nature* is $s_i$ is $p(s_i|x_k)$, which is the updated probability, also called the
      *posterior probability* (posterior to data acquisition); following Bayes' theorem,
      these updated probabilities are determined by

$$p(s_i|x_k) = \frac{p(x_k|s_i)}{p(x_k)} p(s_i) \qquad (7.4)$$

The denominator $p(x_k)$ is the marginal probability of the data $x_k$ and it follows
the total probability theorem:

$$p(x_k) = \sum_{i=1}^{n} p(x_k|s_i)p(s_i) \tag{7.5}$$

Given the data $x_k$, the EUV of alternative j is obtained from the posterior probabilities by

$$E(a_j|x_k) = \sum_{i=1}^{n} u_{ji} p(s_i|x_k) \tag{7.6}$$

For the data $x_k$, the EUV is obtained by maximisation with respect to the alternatives, $a^*$

$$E(a^*|x_k) = \overset{max}{\underset{j}{}} E(a_j|x_k) \tag{7.7}$$

The unconditional maximum EUV (not conditioned to the data outcome) is determined using the EUV for each data $x_k$ multiplied by the marginal probability for each corresponding data $x_k$, e.g.

$$E(a_x^*) = \sum_{k=1}^{r} E(a^*|x_k)p(x_k) \tag{7.8}$$

The subscript $x$ in the alternative $a_x^*$ means that we refer to the data acquisition called $X$.

The VOI assessment is the difference between the EUV for the data acquisition, Eq. 7.8, minus the EUV without data acquisition, Eq. 7.3:

$$VOI = E(a_x^*) - E(a^*) \tag{7.9}$$

In this definition, VOI measures the additional value or utility value resulting from the acquisition of data; when the value of the data is lower than VOI, it is worth acquiring the data; otherwise, it is not worth acquiring the data.

We discuss in Sects. 7.3 and 7.4 two different situations related to data acquisition: whether the data are perfect or imperfect.

## 7.5   Value of Perfect Information

The value of perfect information, VOPI, represents an unrealistic but useful case to understand the value of the data; it is also called the "clairvoyant" case. It assumes that there exists an expert (or tool, or measurement or study) that can accurately predict the future results of a decision problem. In other words, when the data are perfect, for each of the possible data outcomes $x_k$, only one conditional probability fulfils the condition $p(x_k|s_i) \neq 0$ for all $i$: this statement means that for each value obtained from the data acquisition, one *state of nature* is certainly predicted to be true or, said another way, each data indicates only one *state of nature*. However, given one *state of nature*, when the number of data points is larger than the number of states, more than one data value can predict such state, each with a given probability and all adding to 1. Thus, perfect data will always correctly predict the *state of nature*.

For perfect information,

$$p(s_i|x_k) = \begin{cases} 1 \\ 0 \end{cases} \qquad (7.10)$$

Assuming that $x_p$ is the perfect information,

$$E\left(a^*_{x_p}\right) = \sum_{k=1}^{r} E\left(a^*|x_k\right) p(x_k) \qquad (7.11)$$

Finally, the value of perfect information is

$$VOPI = E\left(a^*_{x_p}\right) - E\left(a^*\right) \qquad (7.12)$$

Next, we present an example that will be analysed for VOPI.

**Example 7.2: Value of Perfect Information (VOPI)** Two exploration wells have been drilled in an anticline structure showing good prospects with an average reservoir thickness of 40 ft; seismic interpretation indicates a fault that divides the field into two areas A1 and A2, with approximately 60% in A1.

So far, two wells were drilled in A1, and none were drilled in A2. Using oil–water contact encountered in A1 into A2, the pay thickness is 30 ft. However, the seismic resolution of around 80 ft. does not allow us to identify whether the throw is significant and A2 is deeper than A1 and consequently saturated with water or whether the throw has no displacement between A1 and A2.

With the current information, three potential development scenarios have been prepared, and each one is called a *state of nature*:

(i)   the high (optimistic) scenario, which assumes that the top of both areas, A1 and A2, are at the same level, which means a fault of no significant throw,
(ii)  the medium (average) scenario, which assumes a throw of 20 ft which is half of the reservoir thickness, and
(iii) the low (pessimistic) scenario that assumes the throw of the fault is greater than or equal to 40 ft, which means that block A2 is water saturated.

In all the scenarios, it is assumed there is no communication through the fault, which means the wells are dedicated to produce from block A1 or block A2 but not from both areas. This statement is a strong assumption based on analogues from the area, and it was considered for the financial evaluation of this project.

The production profiles for the three scenarios are estimated, and economics are computed, including the total CAPEX, OPEX, oil prices and taxes and royalties.

The value of the project is calculated in net present value terms. For this example, we use value (monetary value) and not utility (risk is not included in this example).

The decision-maker should decide whether:

(a)    to move forward with the project: sanction the project and start the investment, or

(b)    to reject the project: consequently, lose the investment already made on the cost of the seismic survey and the two wells drilled.

However, before deciding between the alternatives, the decision-maker should also consider acquiring additional data that may impact the assessment of the three states of nature. In this problem, one possible data acquisition action consists of drilling a new well in the area of block A2. This well will indicate whether A2 is deeper than A1 and, if so, the throw of the fault and the fluid saturation in A2. Of course, this data acquisition well will also provide the depth of the oil–water contact in A2.

However, this well will have a cost and delay the project's decision for a few months, which can be totalised as the total cost of the data acquisition.

As discussed previously, perfect data, VOPI, is an ideal case in which the clairvoyant (whether it is data or expert's opinion) predicts the future accurately; for its definition, the VOPI computes the maximum value that the information can provide.

The value of imperfect information (VOII) will always be lower than the VOPI.

We will do this assessment using two different methods: (i) analytical and (ii) a decision tree.

(i)    Analytical

Table 7.2 summarises the initial conditions of this decision problem, e.g., the states of nature and their values and prior probabilities.

Based on information in Table 7.2 and Eqs. (7.2) and (7.3), the expected value for this problem with the current information is $40.0 M.

**Table 7.2** Example 7.2: initial conditions

| State of nature | M$ |
|---|---|
| State high, $s_h$ | 500 |
| State mid, $s_m$ | 20 |
| State low, $s_l$ | −350 |
| Prior probability | |
| Prior high | 0.35 |
| Prior mid | 0.25 |
| Prior low | 0.40 |
| Current investment | −90 |

**Table 7.3** Example 7.2: reliability and posterior probabilities

| Reliability probability | $x_1$ | $x_2$ | $x_3$ |
|---|---|---|---|
| $p(x_k\|s_h)$ | 1.00 | 0.00 | 0.00 |
| $p(x_k\|s_m)$ | 0.00 | 1.00 | 0.00 |
| $p(x_k\|s_l)$ | 0.00 | 0.00 | 1.00 |
| Posterior probability | $x_1$ | $x_2$ | $x_3$ |
| $p(s_h\|x_k)$ | 1.00 | 0.00 | 0.00 |
| $p(s_m\|x_k)$ | 0.00 | 1.00 | 0.00 |
| $p(s_l\|x_k)$ | 0.00 | 0.00 | 1.00 |
| Marginal probability | $x_1$ | $x_2$ | $x_3$ |
| | 0.35 | 0.25 | 0.40 |

**Table 7.4** Example 7.2: expected values for the VOPI

| | $x_1$ | $x_2$ | $x_3$ |
|---|---|---|---|
| $E(a_1)$ | 500.0 | 20.0 | −350.0 |
| $E(a_2)$ | −90.0 | −90.0 | −90.0 |
| $E(a^*\|x_k)$ | 500.0 | 20.0 | −90.0 |
| $E(a^*) = VOPI$ | 144.0 | | |

We will assume that it is possible to acquire perfect information about this problem: perfect information unequivocally identifies the states of nature. Reliability probabilities are estimated based on expert assessment; using Eqs. (7.4) and (7.5), marginal and posterior probabilities are calculated, as shown in Table 7.3.

The expected values of the project, the expected value of each alternative conditioned on the data acquired, the optimum expected value conditioned to data acquired and the optimum unconditional expected value are calculated using Eqs. (7.6), (7.7) and (7.8) and are shown in Table 7.4.

Table 7.4 shows that the value of the project in the case that perfect data are acquired is $144 M. This result means that the project with perfect data has a higher value than the project without data.

The value of perfect information, Eq. 7.9, is,

$$VOPI = \$(144.0 - 40.0)\,M = \$104.0\,M \tag{7.13}$$

If the cost of the data is lower than $104.0 M, it is worth paying for data acquisition; otherwise, the decision should be made based on the current data.

(i) Decision tree

The VOPI assessment can be displayed using a *decision tree*, which is sometimes called a probabilistic tree. The decision tree is used to show the two alternatives available to the decision-maker: (i) decision making based on the current knowledge (also called without data) and (ii) decision making based on new data acquisition (also called with data).

As explained in Chap. 2 of this book, a decision tree is read from left to right. In a decision tree, we have nodes and edges; nodes can be an uncertain or chance node, whose value results from the probabilistic assessment, or decision nodes that assign a value from a decision criterion, which is the maximum between most cases two or more values. In the decision tree, we also have the value or terminal nodes, which are the last to the right of each branch, which are the value of the states of nature. Finally, the edges are the arrows that connect nodes between them.

In the decision tree shown in Fig. 7.2, which corresponds to Example 7.2, the upper branch is the "no data acquisition" scenario, and the lower branch is the "data acquisition" scenario.

On the left side of the "no data acquisition" branch, we have the three values for each scenario of the project and the corresponding prior probabilities; the value of this branch is \$40.0 M, resulting in the maximum between the solution of the three projects, and the rejection alternative. From a financial standpoint, this project produces benefits; here, there is no consideration of risk.

The bottom branch corresponds to the "data acquisition" alternative; here, it is assumed that the data accurately describe what will occur for the value of perfect information. There are three options for the data: it can indicate with certainty that the true state is *high*, *medium* or *low*, so those reliability probabilities are 1 or 0 depending on the conditional probability under evaluation.

In this case, if the data indicate high, there is a 100% chance that the state is high (%) and 0% chance that the state is medium or low; similarly, if the data indicate medium or low, the state is 100% medium or 100% low, correspondingly. These values of reliability probability conduct to probability 1 or 0 for the posterior probabilities, obtained applying Bayes' theorem. This assessment concludes that the value of the project consisting of acquiring the perfect data and making a decision based on the data results has a value of \$144.0 M.

The difference between the two projects is \$(144.0 − 40.0) M = \$104.0 M. This means that if the cost of the data is lower than \$104.0 M, it is better to acquire the data first and then make the decision. However, if the cost of the data is more than \$104.0 M, it is better not to acquire the data and decide based on current information. Therefore, on the shown evaluation, without risk consideration, that decision is to move forward to the project execution.

The value of \$104.0 M is the maximum cost that the decision-maker will be willing to pay to the clairvoyant for the perfect information.

**Example 7.3: Value of Perfect Information** In this second example of the VOPI, the number of states of nature is different than the number of pieces of data (in the previous example, both were equal to three).

This example will be more abstract than Example 7.2 in the sense that we will just mention the states of nature, their values and prior probabilities, which are summarised in Table 7.5.

By making the calculus using Eqs. 7.2 and 7.3, the expected value of this project with the current information is \$5 M.

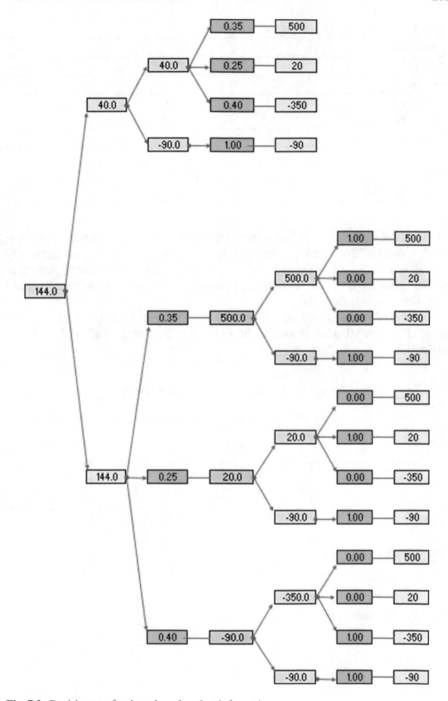

**Fig. 7.2**  Decision tree for the value of perfect information

**Table 7.5** Example 7.3: initial conditions

| State of nature | M$ |
| --- | --- |
| State high, $s_h$ | 1,200 |
| State mid, $s_m$ | 200 |
| State low, $s_l$ | −900 |
| Prior probability | |
| Prior high | 0.30 |
| Prior mid | 0.25 |
| Prior low | 0.45 |
| Current investment | −120 |

To assess the value of gathering new information, we will assume that the data to be acquired have seven possible values $x_1, x_2, x_3, x_4, x_5, x_6, x_7$; the reliability probabilities estimated by the technical experts on the domain related to this particular problem, and the marginal and posterior probabilities calculated using Eqs. (7.4) and (7.5) are shown in Table 7.6.

Based on the information contained in Table 7.6 and Eqs. (7.6)–(7.8), the expected value of this project is calculated, resulting in an expected value of $344 M, as shown in Table 7.7.

**Table 7.6** Example 7.3: reliability, posterior and marginal probabilities

| Reliability probability | $x_1$ | $x_2$ | $x_3$ | $x_4$ | $x_5$ | $x_6$ | $x_7$ |
| --- | --- | --- | --- | --- | --- | --- | --- |
| $p(x_k|s_h)$ | 0.80 | 0.10 | 0.10 | 0.00 | 0.00 | 0.00 | 0.00 |
| $p(x_k|s_m)$ | 0.00 | 0.00 | 0.00 | 0.40 | 0.60 | 0.00 | 0.00 |
| $p(x_k|s_l)$ | 0.00 | 0.00 | 0.00 | 0.00 | 0.00 | 0.50 | 0.50 |
| Posterior probability | $x_1$ | $x_2$ | $x_3$ | $x_4$ | $x_5$ | $x_6$ | $x_7$ |
| $p(s_h|x_k)$ | 1.00 | 1.00 | 1.00 | 0.00 | 0.00 | 0.00 | 0.00 |
| $p(s_m|x_k)$ | 0.00 | 0.00 | 0.00 | 1.00 | 1.00 | 0.00 | 0.00 |
| $p(s_l|x_k)$ | 0.00 | 0.00 | 0.00 | 0.00 | 0.00 | 1.00 | 1.00 |
| Marginal probability | $x_1$ | $x_2$ | $x_3$ | $x_4$ | $x_5$ | $x_6$ | $x_7$ |
| $p(x_k)$ | 0.2400 | 0.0300 | 0.0300 | 0.1000 | 0.1500 | 0.2250 | 0.2250 |

**Table 7.7** Example 7.3: expected values for the VOPI

| | $x_1$ | $x_2$ | $x_3$ | $x_4$ | $x_5$ | $x_6$ | $x_7$ |
| --- | --- | --- | --- | --- | --- | --- | --- |
| $E(a_1)$ | 1,200 | 1,200 | 1,200 | 200 | 120 | −900 | −900 |
| $E(a_2)$ | −120 | −120 | −120 | −120 | −120 | −120 | −120 |
| $E(a^*|x_k)$ | 1,200 | 1,200 | 1,200 | 200 | 120 | −120 | −120 |
| $E(a^*) = VOPI$ | 344 | | | | | | |

Subtracting the value of the project without new information from the value of the project with further information, Eq. 7.9, we get that the value of information is $339 M.

$$VOPI = \$(344 - 5)\,M = \$339\,M \qquad (7.14)$$

This result suggests that the project value can be highly improved through data acquisition. VOPI assessment provides the maximum cost that the decision-maker can invest in data acquisition, bringing value to the project. Suppose the cost of acquiring the perfect data is lower than $339 M. In that case, it is worth acquiring the data because in that case, the project (including the data acquisition) has a higher value than the project without the data acquisition. However, if the cost of the data is higher than the VOPI, acquiring data jeopardises the project.

## 7.6  Value of Imperfect Information

As mentioned in the previous section, in the real world, data are never perfect; the outcomes of the data provide "indications" or suggestions about the uncertain variable that we are investigating, but it will not be able to indicate, without uncertainty, what real state will occur.

This statement is captured, mathematically, by the reliability probability, which, instead of being 1 or 0 as was the case with perfect information, will take any real value included in the interval [0, 1]. e.g. in the case of imperfect information, the conditional probability that the data say the state is high when it is indeed high is less than 1.

**Example 7.4: Value of Imperfect Information** This example is the same as Example 7.2, but now we will consider that the data are imperfect instead of perfect.

The first approach to assessing this project is analytic. Table 7.8 summarises the states of nature, their values and prior probabilities.

**Table 7.8** Example 7.4: initial conditions

| State of nature | M$ |
|---|---|
| State high, $s_h$ | 500 |
| State mid, $s_m$ | 20 |
| State low, $s_l$ | −350 |
| Prior probability | |
| Prior high | 0.35 |
| Prior mid | 0.25 |
| Prior low | 0.40 |
| Current investment | −90 |

**Table 7.9** Example 7.4: reliability and posterior probabilities

| Reliability probability | $x_1$ | $x_2$ | $x_3$ |
|---|---|---|---|
| $p(x_k|s_h)$ | 0.70 | 0.20 | 0.10 |
| $p(x_k|s_m)$ | 0.25 | 0.45 | 0.30 |
| $p(x_k|s_l)$ | 0.20 | 0.30 | 0.50 |
| Posterior probability | $x_1$ | $x_2$ | $x_3$ |
| $p(s_h|x_k)$ | 0.63 | 0.23 | 0.11 |
| $p(s_m|x_k)$ | 0.16 | 0.37 | 0.24 |
| $p(s_l|x_k)$ | 0.21 | 0.40 | 0.65 |
| Marginal probability | $x_1$ | $x_2$ | $x_3$ |
|  | 0.39 | 0.30 | 0.31 |

**Table 7.10** Example 7.4: expected values for the VOII

|  | $x_1$ | $x_2$ | $x_3$ |
|---|---|---|---|
| $E(a_1)$ | 247.1 | −15.7 | −164.5 |
| $E(a_2)$ | −90.0 | −90.0 | −90.0 |
| $E(a^*|x_k)$ | 247.1 | −15.7 | −90.0 |
| $E(a^*) = VOII$ | 63.1 |  |  |

As shown in Example 7.2, the expected value of this problem with the current information is $40 M.

The experts in the domain estimate the reliability probabilities; the posterior and marginal probabilities are calculated using Eqs. (7.4) and (7.5) and shown in Table 7.9.

The reliability probability values are estimated, assuming that the data are imperfect. Finally, the expected values of each alternative, the optimum conditional expected value and the unconditional expected values are summarised in Table 7.10.

Table 7.10 shows that the value of the project in the case that the data are imperfect is $63 M; this means that the project with imperfect data has a higher value than the project without data.

The value of perfect information is,

$$VOPI = \$(63.1 - 40.0)\,M = \$23.1\,M \tag{7.15}$$

If the cost of the data is lower than $23.1 M, it is worth investing in data acquisition; otherwise, the recommended decision is to decide the project's fate using the current data. However, comparing Examples 7.2 and 7.4, we observe a decrease in the value of information, from the perfect data $144.0 M to the imperfect data $63.1 M.

Of course, the "degree of imperfection" of the data is defined by the reliability probabilities. In Example 7.4, we chose a particular set of reliability probabilities, but the value of perfect data will always provide the higher value of data acquisition in all the possible cases.

Examples 7.2 and 7.4 have the same states of nature, prior probabilities and current investment amounts. Still, in Example 7.4, the reliability probabilities are re-estimated to accommodate the uncertainty associated with the accuracy associated with predicting the states of nature by the data.

This problem can also be solved using the technique of a decision tree. Figure 7.3 shows a schematic of this decision tree.

In this case, when the data are imperfect, the value of the project with data acquisition is $63.1 M.

The difference between the projects with data acquisition and without data acquisition is $(63.1 − 40.0) M = $23.1 M. This means that if the cost associated with data acquisition is less than $23.1 M, it is worth acquiring the data first and, based on the result, making the decision. However, suppose the cost of the data is higher than $23.1 M. In that case, it is best to decide without acquiring additional data, which in this case means using the top part of the decision tree where, as observed in Sect. 7.2, it is favourable to move the project forward.

**Example 7.5: Value of Imperfect Information** This example is the same as in Example 7.3, but we are now assuming that the data are imperfect.

Table 7.11 shows the states of nature, their values and prior probabilities.

The value of the project with the current information is $5 M.

To assess the value of gathering information, we will assume that the data to be acquired have seven possible values $x_1, x_2, x_3, x_4, x_5, x_6, x_7$; the reliability probabilities of these values associated with the three states of nature are shown in Table 7.12; also, the posterior probabilities and marginal probabilities calculated using Eqs. 7.4 and 7.5 are included in the same table.

Based on the information contained in Table 7.12, the expected value is calculated, resulting in an expected value of $307 M, as shown in Table 7.13.

Subtracting the value of the project without new information from the value of the project with new information, we get that the value of information is $302 M.

$$VOI = \$(307 - 5)\,M = \$302\,M \qquad (7.16)$$

This result shows a decrease in the value of the data from the case where the data is perfect ($344 M) to the case where the data is imperfect ($307 M). In both cases (perfect or imperfect data), the project with data has a higher value than the project without additional data.

As mentioned before, the perfect data case is unrealistic, but it sets the maximum value that the data can bring to the project. Therefore, if the data's cost is lower than the difference between the perfect data case and the without-new-data case, an assessment of the value of imperfect data must be made considering the impact that the imperfection of the data has on the value of the project.

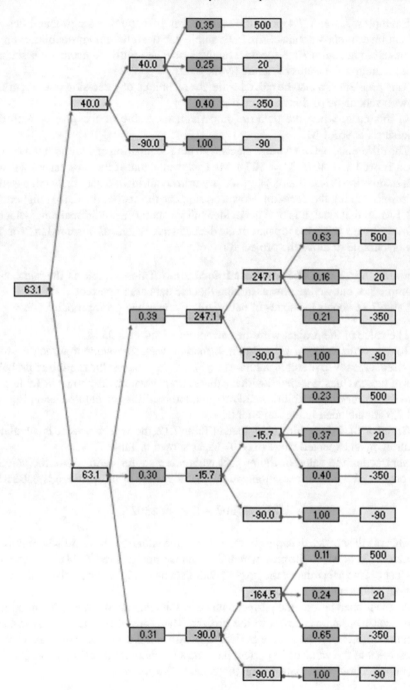

**Fig. 7.3**  Decision tree for the value of imperfect information

**Table 7.11** Example 7.5: initial conditions

| State of nature | M$ |
|---|---|
| State high, $s_h$ | 1,200 |
| State mid, $s_m$ | 200 |
| State low, $s_l$ | −900 |
| Prior probability | |
| Prior high | 0.30 |
| Prior mid | 0.25 |
| Prior low | 0.45 |
| Current investment | −120 |

**Table 7.12** Example 7.5: reliability, posterior and marginal probabilities

| Reliability probability | $x_1$ | $x_2$ | $x_3$ | $x_4$ | $x_5$ | $x_6$ | $x_7$ |
|---|---|---|---|---|---|---|---|
| $p(x_k|s_h)$ | 0.60 | 0.20 | 0.15 | 0.05 | 0.00 | 0.00 | 0.00 |
| $p(x_k|s_m)$ | 0.00 | 0.00 | 0.25 | 0.30 | 0.25 | 0.20 | 0.00 |
| $p(x_k|s_l)$ | 0.00 | 0.00 | 0.00 | 0.05 | 0.15 | 0.25 | 0.55 |
| Posterior probability | $x_1$ | $x_2$ | $x_3$ | $x_4$ | $x_5$ | $x_6$ | $x_7$ |
| $p(s_h|x_k)$ | 1.00 | 1.00 | 0.42 | 0.13 | 0.00 | 0.00 | 0.00 |
| $p(s_m|x_k)$ | 0.00 | 0.00 | 0.58 | 0.67 | 0.48 | 0.31 | 0.00 |
| $p(s_l|x_k)$ | 0.00 | 0.00 | 0.00 | 0.20 | 0.52 | 0.69 | 1.00 |
| Marginal probability | $x_1$ | $x_2$ | $x_3$ | $x_4$ | $x_5$ | $x_6$ | $x_7$ |
| $p(x_k)$ | 0.1800 | 0.0600 | 0.1075 | 0.1125 | 0.1300 | 0.1625 | 0.2475 |

**Table 7.13** Expected values for the VOII

| | $x_1$ | $x_2$ | $x_3$ | $x_4$ | $x_5$ | $x_6$ | $x_7$ |
|---|---|---|---|---|---|---|---|
| $E(a_1)$ | 1,200 | 1,200 | 619 | 113 | −85 | −562 | −900 |
| $E(a_2)$ | −120 | −120 | −120 | −120 | −120 | −120 | −120 |
| $E(a^*|x_k)$ | 1,200 | 1,200 | 619 | 113 | −85 | −120 | −120 |
| $E(a^*) = VOII$ | 307 | | | | | | |

## 7.7 Value of Fuzzy Information

Under this title, we refer to the methodology for comparing the value of incorporating fuzzy and random information into the same framework. This methodology includes fuzzy information and fuzzy outcomes within the classical VOI methodology based on probabilistic and Bayesian theories.

This theory allows us to introduce into the VOI assessment two different sources of uncertainties, the one due to the randomness and partial knowledge and the other one due to fuzziness in the data and outcomes.

Let us assume that we have a set of data values resulting from a data acquisition action $X = \{x_1, x_2, \ldots, x_r\}$; we can define a set of fuzzy events on this information, $\tilde{M} = \left\{\tilde{M}_1, \tilde{M}_2, \ldots, \tilde{M}_n\right\}$, such as "good", "moderate", "poor", "high", "middle" or "low". The fuzzy events will have membership functions $\mu_{\tilde{M}_i}(x_k)$, $k = 1, 2, \ldots, r$ and $i = 1, 2, \ldots, n$, for the data values.

The probability of a fuzzy event $\tilde{M}$ is defined as

$$P\left(\tilde{M}\right) = \sum_{k=1}^{r} \mu_{\tilde{M}}(x_k) p(x_k) \tag{7.17}$$

In this definition, the events are fuzzy because the pieces of data can belong to more than one event, and the degree of belonging is measured through the membership functions.

The membership function defines the degree of belonging that each piece of data has within the events:

$$\mu_{\tilde{M}}(x) : X \to [0, 1] \tag{7.18}$$

In this way, given an event, we associate with each piece of data a real number between 0 and 1 that describes its degree of belonging to that event.

However, when the events are not fuzzy but crisp,

$$\tilde{M} = M \tag{7.19}$$

Then, the membership function can take only two values:

$$\mu_M = \begin{cases} 1, & x_k \in M, \\ 0, & otherwise. \end{cases} \tag{7.20}$$

Consequently,

$$P(M) = \sum_{x_k \in M} p(x_k) \tag{7.21}$$

where $p(x_k)$ considers the marginal probabilities of the data points contained (unequivocally) in each crisp event.

Most of the data that we acquire are inherently fuzzy; e.g. suppose we propose to acquire a core in an oil well to better understand (or reduce) the uncertainty in the porosity values in an area of the reservoir. Assume that the range of possible porosity values in a well, in that referred field, is any real value from 10 to 25%; to assess the value of the information, we cannot, practically, take all the values within such a range. To solve this issue, we create subjective intervals to which a name is assigned. Let us say: the "low porosity value interval", "middle porosity value interval", and "high porosity value interval". This partition on porosity covers the entire range of values.

The question is how to map the porosity values in the range into these intervals, and the answer is by using fuzzy logic.

The intervals defined in this manner are fuzzy because there are porosity values with a partial degree of membership in those intervals. For example, 10% porosity certainty belongs, fully, to the interval "low porosity", but 15% porosity, even though it could belong to "low porosity", could also belong to "middle porosity", so we can say that 15% porosity has partial membership between "low porosity" and "middle porosity". In this case, we cannot use a membership function as described in Eq. 7.20, but we can use a membership function as in Eq. (7.22):

$$\mu_{\tilde{M}}(x_k) = [0, 1], \, x_k \in \tilde{M} \tag{7.22}$$

Equation 7.22 means that the degree of membership for each given porosity value is any number between 0 and 1, where 1 means full membership and 0, not membership at all.

For fuzzy events, instead of looking for posterior probabilities for the data points as we do with crisp data, we estimate the posterior probabilities of the fuzzy events following the same reasoning as with crisp data; then,

$$P\left(s_i|\tilde{M}\right) = \frac{\sum_{k=1}^{r} p(x_k|s_i)\mu_{\tilde{M}}(x_k)p(s_i)}{P\left(\tilde{M}\right)} = \frac{P\left(\tilde{M}|s_i\right)p(s_i)}{P\left(\tilde{M}\right)} \tag{7.23}$$

The reliability probability is

$$P\left(\tilde{M}|s_i\right) = \sum_{k=1}^{r} p(x_k|s_i)\mu_{\tilde{M}}(x_k) \tag{7.24}$$

If we have a set of fuzzy events $\tilde{M} = \left\{\tilde{M}_1, \tilde{M}_2, .., \tilde{M}_n\right\}$, we say the set is orthogonal when

$$\sum_{i=1}^{l} \mu_{\tilde{M}_i}(x_k) = 1, \, \text{for all } x_k \in X \tag{7.25}$$

For orthogonal fuzzy events associated with the new information, the Bayesian approach can be applied.

In the fuzzy formalism, the equations equivalent to 7.6–7.8 are for the expected value of alternative $j$ conditioned to the fuzzy event $\tilde{M}_i$.

$$E\left(a_j|\tilde{M}_i\right) = \sum_{i=1}^{n} u_{ij}\, p(s_i|\tilde{M}_i) \tag{7.26}$$

The expected value of the optimum alternative conditioned to the fuzzy event $\tilde{M}_i$ is

$$E\left(a^*|\tilde{M}_i\right) = \overset{max}{\underset{j}{}}\, E(u_j|\tilde{M}_i) \tag{7.27}$$

The unconditional expected value is,

$$E\left(a_{\tilde{M}}^*\right) = \sum_{i=1}^{l} E\left(a^*|\tilde{M}_i\right) p\left(\tilde{M}_i\right) \tag{7.28}$$

Similar to Eq. 7.9, the value of fuzzy data is,

$$VOI = E\left(a_{\tilde{M}}^*\right) - E\left(a^*\right) \tag{7.29}$$

**Example 7.6: Value of Fuzzy Data** As per Examples 7.3 (VOPI) and 7.5 (VOII), Table 7.14 shows the states of nature, their values and prior probabilities.

The value of the project with the current information is $5 M.

To assess the value of gathering fuzzy information, we will assume that the data to be acquired have seven possible values $x_1, x_2, x_3, x_4, x_5, x_6, x_7$; the reliability probabilities of these values associated with the three states of nature are shown in Table 7.15. The membership functions, defined by the experts and corresponding to the data values and fuzzy set, are also shown in Table 7.15.

The fuzzy reliability probabilities and fuzzy posterior probabilities are calculated using Eqs. 7.23 and 7.24, and the results are shown in Table 7.16.

Finally, the expected value of each alternative and the optimum (maximised) expected value is calculated following Eqs. (7.26), (7.27) and (7.28), and results are shown in Table 7.17.

**Table 7.14** Example 7.6: initial conditions

| State of nature | M$ |
| --- | --- |
| State high, $s_h$ | 1,200 |
| State mid, $s_m$ | 200 |
| State low, $s_l$ | −900 |
| Prior probability | |
| Prior high | 0.30 |
| Prior mid | 0.25 |
| Prior low | 0.45 |
| Current investment | −120 |

**Table 7.15** Example 7.6: reliability probabilities and membership functions

| Reliability probability | $x_1$ | $x_2$ | $x_3$ | $x_4$ | $x_5$ | $x_6$ | $x_7$ |
|---|---|---|---|---|---|---|---|
| $p(x_k\|s_h)$ | 0.60 | 0.20 | 0.15 | 0.05 | 0.00 | 0.00 | 0.00 |
| $p(x_k\|s_h)$ | 0.00 | 0.00 | 0.25 | 0.30 | 0.25 | 0.20 | 0.00 |
| $p(x_k\|s_h)$ | 0.00 | 0.00 | 0.00 | 0.05 | 0.15 | 0.25 | 0.55 |
| Membership function | $x_1$ | $x_2$ | $x_3$ | $x_4$ | $x_5$ | $x_6$ | $x_7$ |
| $\mu_{\widetilde{M1}}(x_k)$ | 1.00 | 1.00 | 0.50 | 0.00 | 0.00 | 0.00 | 0.00 |
| $\mu_{\widetilde{M2}}(x_k)$ | 0.00 | 0.00 | 0.50 | 1.00 | 0.50 | 0.25 | 0.00 |
| $\mu_{\widetilde{M3}}(x_k)$ | 0.00 | 0.00 | 0.00 | 0.00 | 0.50 | 0.75 | 1.00 |
| $p(x_k)$ | 0.18 | 0.06 | 0.11 | 0.11 | 0.13 | 0.16 | 0.25 |

**Table 7.16** Example 7.6: fuzzy reliability probabilities and fuzzy posterior probabilities

| Fuzzy reliability probability | $M_1$ | $M_2$ | $M_3$ |
|---|---|---|---|
| $p(\tilde{M}_x\|s_h)$ | 0.8750 | 0.1250 | 0.0000 |
| $p(\tilde{M}_x\|s_m)$ | 0.1250 | 0.6000 | 0.2750 |
| $p(\tilde{M}_x\|s_l)$ | 0.0000 | 0.1875 | 0.8125 |
| Fuzzy posterior probability | $M_1$ | $M_2$ | $M_3$ |
| $p(s_h\|\tilde{M}_x)$ | 0.8936 | 0.1379 | 0.0000 |
| $p(s_m\|\tilde{M}_x)$ | 0.1064 | 0.5517 | 0.1583 |
| $p(s_l\|\tilde{M}_x)$ | 0.0000 | 0.3103 | 0.8417 |

**Table 7.17** Example 7.6: expected values

| The expected value for alternative and fuzzy set | $M_1$ | $M_2$ | $M_3$ |
|---|---|---|---|
| $E(a_1\|\tilde{M}_x)$ | 1,094 | −3 | −726 |
| $E(a_1\|\tilde{M}_x)$ | −120 | −120 | −120 |
| Optimum conditioned expected value for fuzzy set | | | |
| $E(a^*\|\tilde{M}_x)$ | 1,094 | −3 | −120 |
| Optimum expected value | | | |
| $E(a^*)$ | 268 | | |

# 7.8 Value of Information Applications in the Oil and Gas Industry

VOI is a prescriptive methodology to assess the value that gathering data in the future may add to the present value of a project. VOI gives no value to uncertainty reduction but aims to make the best decision based on the underlying uncertainties. Based on the database of the SPE-related publications, Bratvold et al. (2007) showed that, although the use of systematic qualitative methods in VOI has increased in recent years, it is still far from being a standard application, even when significant investments are involved. VOI is part of the broader Decision Analysis discipline discussed in Chap. 1.

In a project, investment decisions are made at some point in time, and those investments have subsequent monetary consequences. Data gathering is a possible investment decision. Data gathering often has a cost, and it may delay the project with the consequent postponement in project production. When the project's benefits are higher than its harms, it is worth gathering the data. Some relevant papers published in the oil and gas industry that describe the progress on VOI applications are presented below.

Schlaifer (1959), Raiffa and Schlaifer (1961), and Raiffa (1968) developed the fundamental concepts and tools of VOI for applications in business administration; their goal was to enable business administrators to make wiser decisions. They used statistical inference and sampling tools in practical problems of decision making under conditions of uncertainty, where additional information about the state of the world can be obtained through experimentation.

Years earlier, Grayson (1960) published his dissertation (converted into a book the same year), applying the VOI methodology to drilling decisions, where uncertainties are enormous. Grayson's work is, likely, the first reference that uses utility theory and subjective probability theory applied to an oil and gas decision problem. Grayson (1962) used statistical inference terminology to analyse the drilling decision and the value associated with gathering additional information.

Newendorp (1967), in his Ph.D. thesis, discussed the need of developing and including the risk attitude of the decision-maker as part of the VOI assessment. He considered using the exponential utility function in a gas well project subject to several investment options to capture the decision maker's risk attitude. Newendorp explains the logic behind using utility instead of value to capture the decision maker's risk attitude.

Vilela et al. (2017) discuss the difference between implementing the VOI methodology using expected monetary value and expected utility values; a case study related to an oil field development is used to show that in some cases, both approaches conduct to similar conclusions. However, sensitivities on the decision-maker risk attitude can lead to different conclusions.

Newendorp (1972) discussed the logic, mathematical proof, and methodology of Bayes' theorem (developed by Thomas Bayes in 1763), a fundamental mathematical tool behind VOI. He also discussed the concept of sequential sampling or sequential data acquisition. Subsequently, further research and applications expanded the scope of the subject and provided more robustness to the methodology.

Even though Dougherty (1971) did not present any novelty to the theory of VOI, he discussed, in an enjoyable manner, the logic of acquiring new data to improve the project value using the "oilmen" game. He highlights the benefits of using diagrams to represent the decision process, such as decision trees, to ease the visualization and calculate the project value.

Warren (1983) successfully illustrates the value of information method for a project in a frontier area where significant capital investments with high uncertainties are usual. Decision alternatives between completing, dropping, or deferring a project are analysed, and the cost of additional information that exceeds its benefits are estimated.

Lohrenz (1988) discussed the importance of acquiring new information and its role in the decision-making process; he explains the use of the decision trees for projects assessment when new information is obtained. Four petroleum engineering examples are presented where the methodology of the value of information drives the decisions.

Silbergh and Brons (1972) wrote their paper when decision analysis and VOI were still a novelty in the oil and gas industry. They reviewed standard methods of profitability valuation, such as Profit Discounted by the Cost of Money to the Firm, Discounted Cash Flow and Rate of Return, showing the limitations of these methods when valuing uncertain projects, and discussed concepts such as utility functions and VOI. The importance of this paper lies in bringing the development and application of what at that time was a new theory of VOI to real-world problems.

In Moras et al. (1987), a different type of VOI application is discussed. This paper aims to determine the value associated with using several observation wells to monitor underground gas storage reservoir pressure to avoid gas migration and estimate the optimum number of observation wells using reservoir simulation tools. It is one of the first applications found in the literature where VOI and reservoir modelling are used in conjunction.

Gerhardt and Haldorsen (1989) contributed to the use of decision analysis tools, especially VOI, showing in simple examples how these methodologies work, for instance, in a long term well test on a vertical well, a polymer pilot project and in an extended production test on a horizontal well.

Other applications, such as that of Dunn (1992), estimate the value of well logs before running them by describing the uncertainty in the values and predicting the loss due to poor properties estimation. Acquiring logs reduce the uncertainty in well properties almost to zero; the difference between the value of the uncertain values and the values of the "certain" values after acquiring the log provide the value of that data.

Many of the initial developments and applications of VOI in the oil and gas industry were in the subsurface exploration domain. Stibolt and Lehman (1993) discussed the value of seismic information in an exploration asset applied as a European call option, broadening the VOI methodology's scope in the oil and gas industry. Rose (1987) describes exploration activities as a series of investment decisions, whether to develop a project, acquire additional data or modify the hydrocarbon interest sharing.

In the subsurface domain, appraisal activities consist of information-gathering to reduce reservoir uncertainties that may affect the field development. Consequently, VOI has an essential contribution to assessing appraisal activities, which was discussed in detail by Demirmen (1996) for the two types of appraisal activities, screening and optimization. This use of the technique is a significant contribution to VOI, which effectively enlarged the scope of VOI assessment to a broader audience in the oil and gas industry.

Newendorp and Schuyler (2000) developed the fundamental ideas related to VOI from states of nature to Bayesian probability, including examples related to explorations and appraisals activities in oil and gas projects. This outstanding book discusses several key concepts related to VOI and risk attitude, Monte Carlo simulation, and decision-making.

Koninx (2000) develops a straightforward methodology for VOI; he discusses VOI from a methodological perspective, adding examples related to the value of 3D seismic acquisition and appraisal to clearly define the hydrocarbon composition. In other research, Coopersmith and Cunningham (2002) proposed a stepwise methodology to facilitate VOI assessment.

Clemen (1996) in his seminal book explain the VOI methodology using several cases study; he reviews perfect and imperfect VOI workflow and includes an enlightening discussion on risk preferences.

Arild et al. develop the formalism of Monte Carlos VOI that differentiates from the classical VOI. The problem uncertainties are managed either using discrete realizations or Monte Carlo simulation. A case study is used to explain the workflow and compare the two approaches. This paper discusses the advantage that MCVOI brings to the VOI analysis by introducing risk management to the assessment.

Steineder et al. (2018) develop a complete workflow to estimate the value of a pilot test of polymer injection in horizontal well as part of a polymer injection project. The impact of the polymer injected is measured using the Cumulative distribution function of Net Present Value for multiple geological scenarios subject to dynamic modelling.

Vilela et al. (2020a) explore the consequences of making sensitivities in the membership functions into the FIS assessment for a decision problem. The case study presented in Vilela et al. (2019) is used to consider three sensitivities around the original membership functions. More extensive analysis of these sensitivities is recommended to capture the impact in the selection of the membership functions.

Vilela et al. (2020b) develop a holistic methodology to implement VOI; this methodology includes experimental design for screening the most relevant uncertainties in terms of impact on the project's value, which allows determining the data acquisition actions that should be implemented to maximize the benefits. This methodology includes the uncertainty related to the fuzziness in the data to get a proper assessment of the value of the data. The project is assessed using utility functions to capture the risk attitude of the decision-maker. Finally, a FIS is successfully developed to (i) assess the project using two outcome parameters: net present value and internal rate of return, and (ii) capture the fuzziness in the linguistic terms used during the decision-making process.

## 7.9   Sequential Data Acquisition

As mentioned by Bratvold et al. (2007), an essential task for technical people in the oil and gas industry is to produce information that can be used by management in making decisions; this is not peculiar to the oil and gas industry, but it applies to most sectors. One of the most important decisions concerns the acquisition of data that can impact the project uncertainty and value; VOI provides the methodology to assess whether it is worth gathering additional data to benefit the project of interest.

We can consider data acquisition as a sequential process where data is gathered in stages until the decision-maker has the necessary support to decide. This statement can be illustrated using Example 7.2 discussed earlier. Before drilling the first well, a seismic survey is conducted in an area where several oil fields are located, and a structure with a shape of an anticline is found. We do not know if a reservoir exists with only the seismic survey: a volume of rock where hydrocarbon has been trapped and currently remains. A set of possible reservoir volumes can be estimated with the seismic interpretation, but their range of values is huge; properties such as oil saturation, porosity, oil volume factor, and thickness are all uncertain. They are based on values from analogous fields, which provide preliminary figures.

Under this level of information, no decision-maker will decide to sanction the project and commit to developing the field by drilling several wells and building facilities and flowlines.

At this point, the decision-maker approves to drill one exploration well or maybe a set of exploratory and appraisal wells depending on expectations or risk attitude. Each well drilled, each core taken, and each sample acquired are data that can be valued using the VOI approach.

This process of sequential data acquisition which should be based on a VOI assessment will support a better definition of the reservoir by removing or decreasing uncertainties and help the decision-makers in making their decision. Each data point acquired helps better understand the reservoir, reduce uncertainties, and progressively generate a more consistent project with fewer uncertainties. This process will continue until the decision-maker is confronted with a project in which the uncertainties are narrowed down enough, no more data acquisition actions are possible, or the value of the data is higher than the improvement of the value of the project due to uncertainty reduction. At that point, the sequential data acquisition process is complete, and a decision should be made.

## 7.10   Summary

The value of information is one of the central topics of this book. The chapter begins with a discussion about decisions and how uncertainties play a significant role in many decisions. Then, the concept of the value of information is addressed, and its precise meaning is explained. Next, the mathematical formulation of the decision-making and the value of information are developed, and several examples are discussed. The "unrealistic" concept of the value of perfect information is discussed, and an example is presented using equations and graphical representation (decision tree). The more realistic case of imperfect information is developed and presented, and a detailed example is used to explain the methodology step-by-step. Bayes' theorem is used, and the decision tree is included as part of the analysis. In the last part of the chapter, the formalism of fuzzy data is discussed. An example is worked in detail to compare the impact of including the fuzziness in the data in the value of information assessment.

# References

Bratvold, R., Bickel, J., & Lohne, H. (2007). Value of Information in the oil and gas industry: Past, present, and future. *Society of Petroleum Engineers.* In *Proceeding of the SPE Annual Technical Conference and Exhibition*, Anaheim, California, USA, 11–14 November 2007. Paper SPE-110378.

Clemen, R. (1996). *Making hard decisions. An introduction to decision analysis* (2nd ed.). Duxbury Press.

Coopersmith, E., & Cunningham, P. (2002). A practical approach to evaluating the value of information and real options decision in the upstream petroleum industry. *Society of Petroleum Engineers.* In *Proceeding of the SPE Annual Technical Conference and Exhibition*, San Antonio, Texas, USA, 29 September–2 October 2002. Paper SPE-77582.

Demirmen, F. (1996). Use of value of information concept in justification and ranking of subsurface appraisal. *Society of Petroleum Engineers.* In *Proceedings of the Annual Technical Conference and Exhibition,* 6–8 October 1996, Denver, USA. Paper SPE 36631.

Dougherty, E. (1971). The oilman's primer on statistical decision theory. *Society of Petroleum Engineers.* Society of Petroleum Engineers Library (unpublished). Paper SPE 3278.

Dunn, M. (1992). A method to estimate the value of well log information. *Society of Petroleum Engineers.* In *Proceedings of the Annual Technical Conference and Exhibition,* 4–7 October 1992, Washington, D.C., USA. Paper SPE 24672.

Gerhardt, J., & Haldorsen, H. (1989). On the value of information. In *Proceedings of Offshore Europe 89,* Aberdeen, UK. Paper SPE 19291.

Grayson, C. J. (1960). *Decisions under uncertainty. Drilling decisions by oil and gas operators.* Harvard University.

Grayson, C. J. (1962). Bayesian analysis. A new approach to statistical decision-making. In *Proceedings of Economics and Evaluation Symposium,* 15–16 March 1962, Dallas, Texas, USA. Paper SPE 266.

Koninx, J. (2000). Value of information. From cost cutting to value creation. *Society of Petroleum Engineers.* In *Proceedings of the Asia Pacific Oil Conference and Exhibition,* 16–18 October 2000, Brisbane, Australia. Paper SPE 64390.

Lohrenz, J. (1988, April). Net values of our information. *Journal of Petroleum Technology*, 499–503. Paper SPE 16842.

Moras, R., Lesso, W., & Macdonald, R. (1987). Assessing the value of information provided by observation wells in gas storage reservoirs. *Society of Petroleum Engineers.* Society of Petroleum Engineers, Library (unpublished). Paper SPE 17262-MS.

Newendorp, P. (1967). Application of utility theory to drilling investment decisions. Ph.D. Thesis. The University of Oklahoma, USA.

Newendorp, P. (1972). Bayesian analysis—A method for updating risk estimates. *Society of Petroleum Engineers. Journal of Petroleum Technology*, USA, February. Paper SPE 3263-PA.

Newendorp, P., & Schuyler, J. (2000). *Decision analysis for petroleum exploration* (2nd ed.). Planning Press.

Raiffa, H. (1968). *Decision analysis: Introductory lectures on choices under uncertainty.* Addison-Wesley.

Raiffa, H., & Schlaifer, R. (1961). *Applied statistical decision theory.* Harvard University.

Rose, P. (1987). Dealing with risk and uncertainty in exploration: How can we improve? *The American Association of Petroleum Geologists Bulletin, 71*(1), 1–16.

Schlaifer, R. (1959). *Analysis of decisions under uncertainty.* McGraw-Hill.

Silbergh, M., & Brons, F. (1972). Profitability analysis-where are we now? *Journal of Petroleum Technology*, 90–100. *Society of Petroleum Technology.* In *Proceedings of the 45th Annual Fall Meeting*, 4–7 October 1972, Houston, USA. Paper SPE 2994.

Steineder, D., Clemens, T., Osivandi, K., & Thiele, M. (2018). Maximizing value of information of a horizontal polymer pilot under uncertainty. *Society of Petroleum Engineers.* In *Proceeding of the SPE Europec, 80th EAGE Conference and Exhibition*, Copenhagen, Denmark, 11–14 June. Paper SPE-190871-MS.

Stibolt, R., & Lehman, J. (1993). The value of a seismic option. *Society of Petroleum Engineers.* In *Proceedings of the Hydrocarbons Economics and Evaluation Symposium*, 29–30 March 1993, Dallas, Texas, USA. Paper SPE 25821.

Vilela, M., Oluyemi, G., & Petrovski, A. (2017). Value of Information and Risk Preference in Oil and Gas Exploration and Production Projects. *Society of Petroleum Engineers.* In *Proceedings of the SPE Annual Caspian Technical Conference and Exhibition*, Baku, Azerbaijan, 1–3 November. Paper SPE-189044-MS.

Vilela, M., Oluyemi, G., & Petrovski, A. (2019). A fuzzy inference system applied to value of information assessment for oil and gas industry. *Decision Making: Applications in Management and Engineering, 2*(2), 1–18.

Vilela, M., Oluyemi, G., & Petrovski, A. (2020a). Sensitivity analysis applied to fuzzy inference on the value of information in the oil and gas industry. *International Journal of Applied Decision Sciences, 13*(3), 344–362.

Vilela, M., Oluyemi, G., & Petrovski, A. (2020b). A holistic approach to assessment of value of information (VOI) with fuzzy data and decision criteria. *Decision Making: Applications in Management and Engineering, 3*(2), 97–118.

Warren, J. (1983). Development decision: Value of information. *Society of Petroleum Engineers.* In *Proceedings of the Hydrocarbon Economics and Evaluation Symposium of the Society of Petroleum Engineers of AIME*, 3–4 March 1983, Dallas, Texas, USA. Paper SPE 11312.

# Chapter 8
# The Value of Flexibility—Real Options

**Objective**

This chapter explains the logic of applying real options to improve the value of projects, which is one of the central topics of this book. Different categories of real options and methods for implementing real valuations are presented. Justification and workflow of the 'engineering approach' of real options are discussed, and several examples are presented.

## 8.1 Introduction to the Value of Flexibility

The net present value (NPV) is the standard technique for evaluating and demonstrating the attractiveness of capital investment opportunities. If a project has a positive NPV, it will generate benefits for its owner. Consequently, it is worth the capital investment; otherwise, it will produce losses if the NPV is negative and is not worth the investment. If two projects with the same level of uncertainty are being evaluated, the one with the higher NPV turns out to be more attractive to its owner. However, the NPV considers investment as an irreversible, now-or-never decision. Decision-makers can influence project results in the real world, for instance, by abandoning a project if the performance is inferior or expanding it if results are positive. The real options approach captures these possibilities, which the NPV does not assess. In this sense, traditional financial techniques, such as the NPV, internal rate of return (IRR) or value of information (VOI) methodologies, do not consider the complete view of value creation for an investment opportunity.

The real options technique supplements traditional financial methods by considering the value of flexibility (VOF) in the decision-making process. However, compared to more conventional approaches, real options require a more sophisticated understanding of the financial theory and more time and resources for analysis; this complexity in applying real options is the main reason that it has become the primary valuation method.

© The Author(s), under exclusive license to Springer Nature Switzerland AG 2022    229
M. J. Vilela and G. F. Oluyemi, *Value of Information and Flexibility*,
Petroleum Engineering, https://doi.org/10.1007/978-3-030-86989-2_8

Begg and Bratvold (2002) discussed one crucial question: what is the best way to deal with uncertainty: pay to reduce it (VOI) and/or plan to manage the consequences derived from it (VOF)? There is no one answer to suit all problems but one answer for each issue. The same authors suggested four cases in which flexibility can add value:

(A)   Where it is not possible to acquire new information to reduce uncertainty.
(B)   When the value increase due to flexibility is more significant than the increase due to data acquisition.
(C)   When the uncertainty remains after data are acquired.
(D)   When flexibility creates value.

A financial option is a right, without the obligation, to buy or sell an underlying asset within a given period and at a determined price. A commercial option derives its value from (i) the underlying value of the asset, which often fluctuates during the period of the opportunity, and (ii) the decision made by the option's holder to exercise or hold it.

Myers (1984) introduced the real options methodology to strategic corporate planning based on the similarity between a financial option and an investment project. The fundamental justification for using real options arises when, due to the uncertainty in a project, an option is identified as benefiting from the upside opportunities while avoiding the downside risk. A real option is a right that the decision-maker (option holder) has to make, change or reject an investment opportunity.

The real options method is rooted in the pricing of financial stock options, which are contracts sold against a certain premium, giving the buyer the right, but not the obligation, to buy a stock at a predetermined price. Real options have a defined expiry time, similar to when the financing options can be executed. The volatility captures the cash flow variability.

A real options analysis (ROA) assesses the value of flexibility in the decision-making process, such as the value associated with making decisions that partially or totally change a project. Flexibility allows the uncertainty to be managed, and different options have different values, so if there is no operational flexibility, there are no real options. Wherever there is the possibility to alter a project's course, there is an opportunity to create value. Real options help the decision-maker manage uncertainty by removing the more negative outcomes and keeping the most favourable ones.

ROA explicitly recognises the evaluation process and the potential benefits of future modifications in the configuration and operations in the system at a future time; by including these future decisions in the valuation analysis, the value of the project can be increased, allowing it to respond efficiently to the changes in the world. In this sense, ROA provides a bridge between financial theory and strategic planning.

In the traditional valuation of projects, two complementary methods are used: quantitative, such as the NPV or IRR, and qualitative, based on strategic elements; these two general methods can produce contradictory recommendations. Additionally, in many cases, the traditional approach to valuation does not satisfy managers' strategic consideration, suggesting a more insightful analysis. Real options can provide some solutions in this situation.

These methods do not consider the variables' uncertainty that may exist in the future. In oil field developments, investment decisions are taken assuming the fluctuations in a given oil price; they also follow a determined line of development and incorporate the probability of failure into the project's expected value without considering the benefits due to a possible flexible line of development.

*Operational flexibility*, which can positively or negatively impact the project execution, is not captured by the traditional economic valuation methods, which consider only one future scenario. ROA is a complementary tool to the standard economic assessments that involve a systematic and methodical analysis of a project's strategic components. As Trigeorgis and Mason (1987) mentioned, ROA provides projects' decision assessment with flexibility. As Coopersmith and Cunningham (2002) noted, many people find that the value of information and real options is complex or cumbersome to implement in the oil and gas industry, which explains why they are rarely used, especially when results are required within time limits.

Real options capture the values associated with a project that is not commonly built into the NPV calculation, such as:

(i)   The value that developing a project can provide for other future projects or growing opportunities; in this sense, the complete evaluation of a project should not be restricted to the potential cash flow that it generates but should also include the potential to produce information, data, and knowledge for future projects. In the oil and gas industry, the value of a project in an unknown area, such as the Arctic region or the unconventional oil in the US, can support future developments in those areas or similar ones in terms of knowledge and expertise mention two particular aspects.

(ii)  In some cases, there is a value associated with flexibility in selecting materials and equipment or options for exiting the complete project or partial disinvestment; this type of option has a higher value as the project uncertainty is higher.

(iii) Another type of real options, critical in this book, is associated with project development's sequential investment. This type of real options consists of the expenditure related to one milestone; then, another sequential investment can be made depending on the outcome. This type of real options is linked to the sequential value of the information approach, as discussed in Chap. 7.

## 8.2   A Brief Explanation of Valuation

The standard tools for making financial decisions are the NPV, internal rate of return (IRR), payback period, and return on investment (ROI). The NPV is the difference between the discounted cash flows expected from the investment and the initial investment, as described in Eq. (8.1):

$$NPV = \sum_{t=0}^{N} \frac{C_t}{(1+i)^t} \tag{8.1}$$

where $t$ is each period of evaluation, $N$ is the total time of the project, $i$ is the discount rate and $C_t$ is the cash flow at time $t$. The discount rate is the interest at which the company can borrow money from financial entities. A project should be accepted if it has a positive NPV and rejected otherwise, the better the NPV, the better the project. The IRR is the discount rate that makes the NPV of the project zero; hence, the IRR is the value of $i$, which makes Eq. (8.1) equal to zero. The higher the IRR, the better the finance of the project.

The project's payback is the time required for a return on the initial investment; in other words, the payback is the length of time that investment takes to reach a break-even point. The shorter the payback period, the more attractive the investment. The ROI is a comparison of the money gained or lost with the amount of money invested; it is estimated using Eq. (8.2):

$$ROI = \frac{V_f - I_i}{I_i} = \frac{V_f}{I_i} - 1 \tag{8.2}$$

where $V_f$ is the final value and $I_i$ is the initial investment.

For $ROI > 0$, the investment is profitable.
For $ROI < 0$, the investment is a loss.

These methods for estimating a project's value do not consider their uncertainty, either technical or economic. For example, when using these estimations in an oil project, no consideration is made of the original oil in place (OOIP), well producibility or oil price uncertainty; besides, these methods consider only one line of development for the future of the oil field. Real options analysis considers uncertainty and flexibility as part of the decision-making process. In the financial stock market, there are two types of options:

A.  *Call options*, in which the owner of the contract (the buyer) holds the right, but not the obligation, to buy a stock at a predetermined price (the strike price) within a specific time (the exercise date or expiration date);
B.  *Put options* in which the contract owner holds the right, but not the obligation, to sell a stock at a predetermined price and within a specific time.

The option holder has the right, but not the obligation, to take action to buy (call option) or sell (put option) something within a specific time limit, now or in the future, for a predetermined price. There are two types of financial options: (i) *European options*, which can only be exercised at the maturity date, and (ii) *American options*, which can be exercised at any time until the maturity date.

Independently of whether real options are of the European or the American type, they are classified into different categories depending on what they bring to a project. In this book, we present the following categories of real options:

(1)   *Defer/learn*: The defer option gives its holder the possibility of putting the decision off for a specified period. The objective is to gain more time to reduce uncertainty reduction, perhaps through data acquisition or stabilisation of the system, uncertain business reduction or favourable market conditions. This option is structured as a call option. Its value may come from a decrease in cost, a change in in-laws, the arrival of new technologies, the expectation of an oil price increase or more information that reduces uncertainty. If the project can be developed in a phased manner, the initial phases can be used to learn. Based on the learning, the project can be modified to optimise the total project returns.

(2)   *Investment/growth/extension*: The investment option gives its holder the possibility to acquire, in the future, an additional part of the project, a module or even another asset leveraged on the original asset to increase the output of the project; this option includes staged investment opportunities and projects developed for phases in which the first stage indicates the possibility of expanding the second stage and so on. This option is similar to a financial call option, in which the owner has the right to scale up a project for a fixed price, and it is the most common type of real options. Using this option, the decision-maker can progress with the first project even if expected to have a negative return.

(3)   *Alter the operating scale*: This option gives its holder the possibility of increasing the output level for the same asset with an investment to scale up the production capabilities; incorporating flexibility into the design to react to uncertainty in the future can have a higher value than the value based on traditional value analysis. This option refers to the flexibility built into the initial project design to react to uncertainty in the future, which provides higher value than the traditional valuation methods.

(4)   *Disinvest/reduce/abandon*: The disinvestment option gives its holder the possibility to reduce the investment size or even abandon it at specific times during the project's life. This option allows the holder to scale back the project and reduce the overall operation, and it is structured as a put option; in the case of abandonment, it gives the holder the right to exit the project. This option applies especially to long-term projects in which uncertainty in the economic situation, low monetary returns and technical issues recommend reducing or abandoning a project to prevent significant losses.

(5)   *Switch*: This option allows processes and products to be switched when necessary due to their cost, availability, or prices.

(6)   *Extend*: This option gives its holder the right to lengthen the time of an asset's expiration, and it is a call option.

In most valuation problems, projects have more than two of the above real options simultaneously.

As Amram and Kulatilaka (1999) mentioned, in the traditional approach to investment based on net present estimates, the higher the uncertainty, the lower the value of the project. They defined real options as an extension of the theory of financial options to real assets; however, the real options approach shows that higher uncertainty creates value as options are identified and used to respond with flexibility to the system or market. An increase in uncertainty leads to a rise in asset value. Flexibility allows the decision-maker to change the course of a project as new information arrives.

One graphical representation frequently used in real options by financial practitioners is the value-at-risk-and-gain (VARG) plot; it depicts the cumulative distributions of outcomes resulting from the parameters' uncertainty. The VARG indicates the percentage of outcomes that is above or below any specific cut-off. This graphical representation is frequently used to emphasise the likelihood of losses occurring and to suggest possible gains. In real options analysis, the interest lies in identifying the scenarios that produce losses, finding ways through flexible designs, reducing the downsize cases and identifying the situations that make the maximum gains to increase the upsize cases. Figure 8.1 shows the VARG when the project analysis is changed from standard financial to flexible project analysis.

There are three major approaches to valuing real options:

1.  *The partial differential equation approach*: As its name suggests, this approach is based on mathematical techniques. It has many applications, and it is well represented by the *Black–Scholes option pricing model* developed in 1973 by Fisher Black, Robert Merton, and Myron Scholes. It is based on solving partial differential equations and can only be implemented for European options because it needs a fixed decision date. It cannot be used with dividend payments and compound options. It is simple to use because it depends on six input variables in an equation. This approach calculates option values by equating the change in option values to tracking portfolio values. However, some of the assumptions

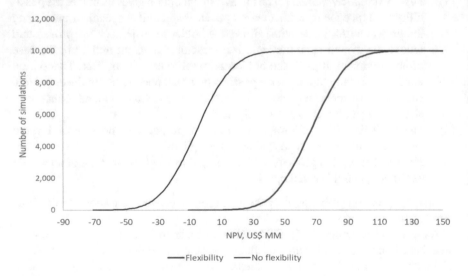

**Fig. 8.1** VARG plot

are not easy to estimate, and the method works as a "black box", which makes its application to real-world problems challenging and infrequent. This method will not be discussed further due to its limited applicability to the scope of this book.

2. *The dynamic programming approach*: This approach extends a project's possible values through the option's duration. Due to the limitations observed for the partial differential equation approach, researchers have created another approach to valuing real options that overcome its weaknesses. The most frequently used method in the dynamic programming approach is the *binomial lattice method* developed by Cox et al. (1979). It allows users to search for the optimal strategy given the decision made in the previous period and discount the value of the best strategy to time zero in a backward process. The binomial lattice method can be used for European and American options. This approach creates binomial trees, making it very descriptive, providing the real options values in each step and illustrating the intermediate decision-making process between each step and the option's expiration time. It is very often used for real options applications, and it assumes that starting from the initial value $X$, the value in the first period can move either upwards or downwards, corresponding to an increase in $u$ with a probability $p(Xu)$ or a decrease in $d$ with a probability $p(Xd)$. In the second period, it will move similarly so that, at the end of the second period, the possible values are $Xu^2$, $Xud$ and $Xd^2$; this process continues in a sequence of periods with such binomial movement, each representing all the possible values that the assets produce during the option's life. This characteristic of the binomial option shows the intermediate decision-making processes between now and the expiration time, allowing users to understand how a decision should be made at each point in time. This method applies only to one type of uncertainty, so we will not discuss it in more detail in this book.

3. *The simulation approach*: This approach extends a project's value based on many possible scenarios from the present to the option expiration time. This approach's most frequently used method is the Monte Carlo method, which simulates the project from the estimated input variables. This method, which can handle several types of uncertainty with values generated using probability distributions, calculates the option's value by randomly simulating many possible future scenarios for the uncertain variables. It also enables us to handle path-dependent real options, offering a significant advantage. For the Monte Carlo simulation method to be comparable with the Black–Scholes equation or the binomial lattice methods, it must be risk-neutral, making the Monte Carlo method complex and reducing transparency.

## 8.3  Applying Real Options to Real Problems

Real-world problems usually are complex, making it challenging to consider all forms of uncertainty correctly. In the plan for developing an oil field, hundreds of designs can be used to assess the project's value.

On the subsurface, different assumptions of the reservoir properties are made to accommodate the uncertainty (porosity, permeability, saturation, oil in place, etc.), which represent the actual values within ranges of uncertainty (data are acquired in wells and extrapolated/interpolated to the entire field). Wells' productivity forecast is based on assumptions subject to interpretations, and different well configurations can be used. The wells' completion and production configuration (natural flow, artificial lift, open or cased hole, etc.) have consequences for the well and field deliverability.

There are several alternatives for flowline configurations on the surface, wells connected separately to a manifold and then to a shared facility or independent well connections, the size and type of flowlines, flow assurance, different chemical requirements injections and many more. Commercial aspects are also uncertain, including the oil price and the various contract options. These elements are a few of the considerations that need to be made when designing a plan to develop an oil field.

In general, it is overly complex to estimate the right combination of factors to produce the most reliable outcome. The standard approach to evaluating a project is to use, for each parameter, what experts in their field (petrophysicists, geologists, drillers, completion engineers, facilities engineers, cost experts, financial and market experts, etc.) assume to be the *best estimate*, sometimes called the *most likely values*. Sometimes, two or three scenarios are built to consider variations in some of the parameters that drive the project. Still, the uncertainty and opportunities derived from them are hardly found in projects' value analyses.

The standard approach to project valuation can be improved by including flexibility in the project, acknowledging the uncertainty, and designing the project to capture the opportunities derived from it. In this way, the decision-maker can react to changes occurring during the project's development. Capturing the value of flexibility starts with recognising uncertain project scenarios upfront and then designing the project to adapt to the changing circumstances during its lifetime.

To prepare for the unknown future, we can consider the process *"if/then/else"*: if something happens, then a decision is taken; otherwise (else), a different decision is taken because the original condition is not fulfilled. This type of evaluation cannot be made using standard evaluation but requires a real options evaluation. The common assessment assumes that the first investment is made, and, from that, the project will move in one direction without intervention and changes, which is hardly the case in the real world. Real options assume that the only alternative for a project's progress is not necessarily the initial stop/go decision but a sequence of little steps with decisions between them depending on the project's performance and the evolution of the external circumstances.

In 2001, Copeland and Antikarov stated that, in 10 years, real options would replace the NPV as the tool to make investment decisions; we know that this prediction failed. Later, in 2008, Willigers and Bratvold indicated that real options had not been applied as successfully to real-world problems as initially expected, and the main reasons seem to be the following:

(i) limitations in the valuation techniques, making it challenging to model and value real-world problems, and

(ii) complexity in the valuation techniques, making it difficult to understand and apply them, requiring profound knowledge of a specific domain and financial expertise; meeting these requirements is challenging. It is not easy for people to become familiar with the real options techniques.

The financial approach to real options is not the best alternative when dealing with real assets, as problems are usually relatively complex to solve using closed-form equations. Financial terminology exacerbates engineering problems.

## 8.4 Real Case Applications of Real Options—Flexibility

Flexibility has been used in many different domains. Binder et al. (2017) discussed the design of a hybrid energy system and showed that, in this case, adding flexibility to the design to allow future modifications of the configuration to respond to potential economic and technological changes increased the value of the project.

There are compelling examples of the use of flexibility to gain a project's upsides due to uncertainty. In real cases, Neufville et al. (2006) developed the engineering-based method for real options, showing its application in the design of infrastructure systems, notably a multistorey car park structure of several floors. In this example, the objective was to design a garage structure of several floors to cope with future demand. The initial NPV assessment indicated that the optimum design had six floors; however, when the future demand's uncertainty (or volatility) is included in the Monte Carlo simulation assessment, the analysis suggested that the five-floor design had a higher expected value. At this stage of the analysis, the authors suggested that a source of flexibility was represented by building the structure with sufficient capacity to support several floors. Initially, they had just the number of floors that remove the more negative cases and enable the manager to decide later, during the life of the garage, to build possible extensions (more floors) if the demand increases. This design would provide the project with flexibility in the future. Still, it would entail an additional cost, in the present, for building big columns to support a potential increase in floors. In this case, the optimum design consisted of a garage of four floors with the potentiality to expand to six floors. This design proved to have a higher value than the ones based on NPV estimation.

Begg et al. (2004) showed how fluctuations in the oil price could be used to extend the life of an oil field beyond what is recommended by a standard discounted cash flow analysis. They also showed how oil production could be optimised by shutting in or choking back oil production following the oil price fluctuation and avoiding total oil production, with a negative NPV, in periods with a low oil price. Real options for capturing the value of flexibility have also been illustrated for production optimisation, for example, in the implementation of intelligent well completions (Han, 2003), to evaluate the benefits obtained by an oil field development in which, instead of using conventional well completion, an intelligent system is applied to allow flexible operation for the monitoring and control of fluid production and injection by zone in real-time. Dezen and Morooka (2002) analysed, qualitatively, the benefits that flexibility and real options can bring to the development of an offshore field and highlighted how different forms of flexibility could be implemented.

Real options have also been applied in areas such as renewable energy. Kim et al. (2016) discussed the real options analysis of photovoltaic projects to assess the impact of climate change on solar energy production. The climate-related risks in solar power are temperature and isolation changes, and the uncertainty of those factors for the project's value is captured through the real options analysis.

Jafarizadeh and Bratvold (2009) define flexibility in a project as the ability to change the course of action in the future. This concept is fundamental to obtaining benefits from the project's uncertainties; they describe different real options applicable to oil and gas investments, showing that the project's value improves by applying this valuation method.

Muller (2018) evaluated dairy farmers' investment decisions in the Netherlands subject to several institutional risks using the real options technique. These institutional risks (new laws or rules) profoundly impacted the standard evaluation of dairy projects that involve significant investment in cattle acquisition, the building of stables, extra land, feedstock, and so on. While typical evaluation results in negative cases, adding flexibility to the evaluation results in attractive investments.

Babajide et al. (2009) discussed an example of an offshore development located in the Gulf of Mexico consisting of two fields that can be developed separately or jointly; the alternatives were to build just one of the two fields or to build both together. For simplicity, in the case study, the uncertainty considered was only associated with the OOIP. The base case showed that joint development had a higher value. When the value of flexibility, consisting of the capability to expand the number of wells after five years of production depending on the base case results, was evaluated, this flexible design showed a better NPV. This case study demonstrated that flexibility improved the value of the project substantially.

Willigers and Bratvold (2008) discussed applying the Longstaff and Schwartz approach to real options, called the least-squares Monte Carlos method. First, they provided an example, already presented by Brandao et al. (2005): a project consisting of the exploitation of an oil field for ten years, in which a deterministic evaluation was made with two uncertain variables, the operating cost and the oil price, which were evaluated using Monte Carlo simulation. After the fifth year, three decision options were considered, which may change the project's course: divest for a price, buy a current partner's share, or continue the project's duration. The first two alternatives can be evaluated using real options, and they are an example of flexibility. A least-squares regression was generated for performing the evaluation that yielded the NPV as a linear function of the cost and the oil price. It was demonstrated that, depending on the uncertain variables' value, real options valuation suggests the best way to proceed. A more elaborate example of a gas producer's decision between a long-term gas contract and trading gas in the market was also discussed. Three variables were considered to be uncertain: the gas price, cost and production. In this example, it was proved that the optimum decision was the trading of gas when the gas and the cost were considered to be uncorrelated variables, and the less-preferred project was the long-term contract.

## 8.5   The Engineering Approach to Real Options Valuation

The engineering-based approach avoids the complexity of financial theory and focuses on the value of flexibility in engineering systems. In the engineering approach to real options, three main elements are identified:

a.  *The design elements*, which are a constituent part of the system. In the development of an oil field, the design elements are wells (horizontal or vertical wells), production facilities, water injection facilities' configurations, production methods (gas lift, electric submersible pump, etc.), flowlines and others.

b.  *The management decision rules*, which represent the procedures to manage and operate the system. In an oil field development, these are the artificial lift system's working conditions, the injection pressure for water injection wells, the manifold pressure, the selected wells' architecture, etc.

c.  *The uncertain variables*, which are the variables that we do not know with certainty and of which the values are beyond our control. For example, in an oil field development, they are represented by the OOIP, the well deliverability, the reservoir pressure response to the production, the oil price, etc. Each manifestation of an uncertain variable over the project's life generates one uncertain variable scenario. For example, if we have three possible values for the OOIP at the beginning of oil development, each corresponds to a scenario of this variable; similarly, a 20-year forecast of the oil price or oil production is a scenario of these variables.

The decision-maker should select the combination of design elements and management decision rules that generates the system's optimum value and performance. The system has uncertainty in several input variables: the oil price, OOIP, well productivity, cost of materials, and equipment. This uncertainty affects the system's value and performance and is outside the control of the decision-maker. Design configurations and management decision rules perform exceptionally well in one uncertainty setting but very poorly in the other one.

For that reason, for each possible uncertainty scenario (value), the analyst should create optimum design elements and management decision rules. In practice, that problem is intractable due to the massive number of possible future scenarios for the uncertain variables. When flexibility in the system is added to the analysis, the difficulty is even greater. Accordingly, the designer should consider a limited set of representative or "relevant" scenarios to define the optimum design elements and management decision rules for each one.

We can select a limited number of "relevant scenarios" for the variables that capture practical implementation uncertainty. These relevant scenarios depend strongly on the type of variable. If we refer to a variable such as the OOIP, experts estimate their current understanding of that uncertainty by calculating the worst, most likely and best scenarios; in this situation, a decision tree is the best tool to perform the uncertainty analysis. Concerning a variable such as the oil price, experts typically use Monte Carlo simulation to generate many future scenarios (thousands, millions) for the oil price over time. These many scenarios can later be classified into just a few "relevant scenarios" using sound rules. For example, all the oil price profiles with a low initial value and a low decline are one "relevant scenario". All the oil price profiles with a high initial value and a low decline correspond to another "relevant scenario" and so on. In this way, for the oil price, we end up with four "relevant scenarios". Each of the Monte Carlo simulation runs would be associated with one of these relevant scenarios. The relevant scenarios created for the Monte Carlo simulation runs are similar to those selected in decision tree analysis; in this example, we end up with four relevant scenarios.

However, the relevant scenarios are created. They are very limited in number. We determine the optimum design elements and management decision rules that optimise the system's value and performance for each one. Each set of design elements, management decision rules and relevant scenarios is called an operation plan. The complete set of operational plans is called the *catalogue of operation plans*. An operating plan consists of the management and operation of a system that joins together a specified set of design elements and management decision rules within a set of uncertain variable scenarios. Given different uncertain scenarios, we can create other design elements and management decision rules that optimise the system's value and performance; the design of the experimental techniques can define the design elements and management decision rules that optimise the system. In that sense, the different design elements or management decision rules can be considered different levels of the experimental variable. The design of experiments (DOE) allows the user to select the appropriate experiments to choose the optimum operation plan given an uncertain scenario. Wang proposed the use of DOE to obtain the optimum design elements (Wang, 2007).

The workflow for the engineering approach consists of four steps:

(i)    *Build an initial model of the system to measure value and performance.* Here, the design elements, management decision rules, uncertainty, and a metric for evaluating value and performance are identified. In this step, the base case or most likely scenario is developed; a deterministic assessment of the system is undertaken without considering either uncertainty or flexibility.

(ii) *Explore the system's uncertainty, propose a limited set of uncertain variable scenarios and assess their impact on the model by calculating the distribution of outcomes that can occur from the uncertainty.* The effect of the system's uncertainty can be estimated with the following: (A) Using decision trees for modelling alternative options for a system in a set of uncertain scenarios, each one with one value for the uncertain input variable and the corresponding probability; this evaluation permits the user to compare the benefits of several options and choose the one with the higher NPV, and (B) Using Monte Carlo simulation to estimate the uncertainty in NPVs due to the uncertainty in the input variables; many (thousands, millions) different scenarios are generated, which are later reduced using the criteria outlined in the previous section to develop the "relevant scenarios". A catalogue of operation plans is built consisting of "relevant scenarios" representing the uncertainty observed in the many (thousands, millions) different scenarios created in the Monte Carlo simulation plus the corresponding design elements and management decision rules that optimise the scenarios. This Monte Carlo analysis allows the estimation of the downsize cases' likelihood and the reasons behind them. The selection of the method to be used, either the decision tree or the Monte Carlo simulation, depends on the nature of the uncertainty and the problem's characteristics. In either case, the results can be visualised using VARG plots (discussed below).

(iii) *Analyse and identify the primary sources of flexibility in the system and include them in the model.* Including flexibility in the model allows the system to adapt to the uncertainty, thus limiting the downside risk and taking advantage of the upside potential. The flexibility is incorporated directly into the decision tree or Monte Carlo simulation. The analysis is usually conducted using the VARG plot and experts' knowledge to identify the modification that should be made in the base case design to allow the removal or mitigation of the downside cases and take advantage of or promote the upside cases, assess the value of the project with the flexibility incorporated into it and estimate the corresponding NPV. In this step, we should also define the decision rules for exercising the design options. There is not yet an established method for identifying flexibility.

(iv) *Select the flexibility with the highest values and compare it with the base case (no flexibility) using the expected value to assess the value of adding flexibility to the system.* The selection of the flexibility that adds more value to the system is made by comparing the flexible case with the inflexible uncertain case. The VARG plot indicates the expected value and risk reduction for the flexible case due to the system's implementation of flexibility.

This process for valuing a project's flexibility can be improved as more applications, experience, and knowledge are gained.

## 8.6    An Example of Flexibility in the Oil and Gas Industry

To make a real options assessment, we should understand the options that give project flexibility. The decision-maker should manage to optimise the project's benefits; these options are then modelled, and their value is assessed. Thinking creatively to find alternative paths to develop a project is essential to implement flexibility in a project. For example, splitting the project and the expenditures into phases, depending on specific decisions based on the response occurring in each stage, is an example of adding flexibility to a project. Each phase depends on the previous results, allowing the project to move in the direction that optimises the benefits.

An offshore oil field has been discovered and appraised with six wells; based on the well test results, preliminary production forecasts have been generated for the 15-year project duration. However, there is uncertainty regarding the strength of the aquifer support for oil production.

(i)   Base case

The first case for this project assumes that the production will be carried out without implementing water injection. The project's assessment depends on the scenarios that the technical experts predict for future development without external support. In this case, the experts generate high, medium and low scenarios, which corresponds to strong, medium and weak aquifer support. These forecasts are shown in Fig. 8.2.

In all the scenarios, oil production starts in the year 2021, and during the first two years of production, the forecasts are the same for the three scenarios, when the primary drive mechanism is oil expansion. However, after two years, the three forecasts are differentiated depending on the aquifer support predicted, yielding the high, medium, and low forecasts.

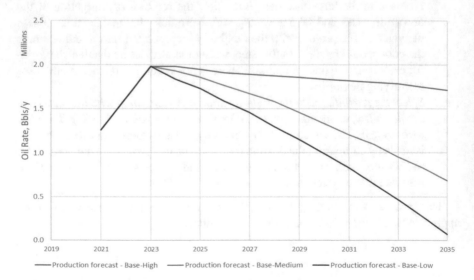

**Fig. 8.2**  High, medium, and low production profiles

Figure 8.2 indicates that, based on the knowledge and experience of the technical team involved in this project, if there is a strong aquifer in this field, the project will be able to deliver a high profile. However, if there is a medium aquifer, it will deliver the medium forecast. Finally, if the aquifer is weak, the project will provide a low forecast; we assume the same numbers of wells in the three cases.

The oil price is assumed, in all the scenarios, to be US$58 per barrel during the year 2021, with an annual increase of 1%. The OPEX cost is assumed to be $1.0/Bbls, based on figures from nearby fields' operations. For this exercise, the project is evaluated as a whole, excluding taxes, royalties, and any other form of penalties. The total CAPEX of this project, excluding facilities for water injection, is $710 M. Figure 8.3 shows the project's cash flow when the production starts, one year after the investment.

Table 8.1 shows the NPV, the estimated probability of each scenario and the expected value of this project, $-19.4 M.

This negative expected value suggests stopping the project. Additionally, the low case has an exceptionally damaging risk because of the significant likelihood (30%) of losing $107.8 M. This low case, with a high chance of losing that significant amount of money, must be improved because the company does not sanction a

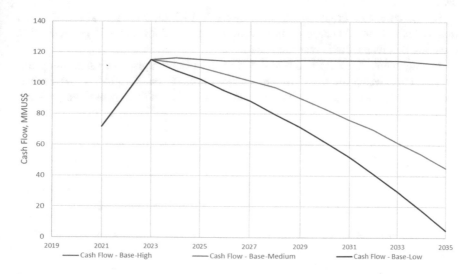

**Fig. 8.3** The cash flow of the project from the first oil for the base case

**Table 8.1** Base case financial assessment

| Scenario | NPV ($ M) | Probability (%) |
|---|---|---|
| High | 106.3 | 20 |
| Medium | −16.6 | 50 |
| Low | −107.8 | 30 |
| Expected value ($ M) | −19.4 | |

project with such risk. Based on previous results, management is willing to search for alternative development to improve this project's financial outcome. The technical team proposes implementing the water injection to reduce the drop in oil production observed in the low and medium cases.

(ii)   Water injection case

Consideration is made of the alternative to build, upfront, the facilities to inject water, reducing the risk associated with the aquifer's strength. The water injection facilities have an additional cost of $100.0 M. In this case, the water injection facilities are designed to ensure that the high oil production forecast scenario will always be met.

From an economic standpoint, there is no significant difference in the facilities' cost to support the production in the low and medium cases, corresponding to the weak and medium aquifer support scenarios. In the water injection case, because the water injection will provide the necessary support for oil production, the three production profiles collapse in the high case, as shown in Fig. 8.4.

Figure 8.5 shows the corresponding cash flow starting with the first oil; of course, all the cash flows are the same.

Table 8.2 shows the NPV, probabilities and expected value of the water injection case.

In this case, this project's expected value is $6.3 M, which is $25.7 M better than the base case. This case project improves the NPV turning on a positive NPV, meaning that its pursuit is recommended.

Management still considers that the NPV should be higher for sanctioning this project, considering the high investment and the risks that the project carries. The technical team propose a flexible alternative as a means to improve the project's value.

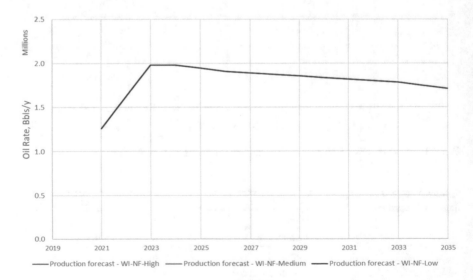

**Fig. 8.4**  Production profiles with water injection facilities built upfront

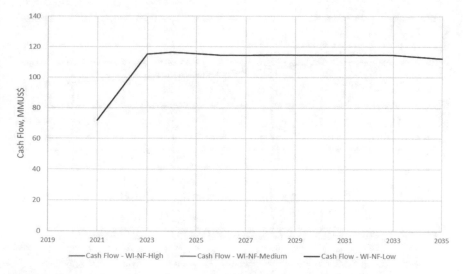

**Fig. 8.5**  The cash flow of the project from the first oil for the water injection case

**Table 8.2**  Financial assessment of the water injection case

| Scenario | NPV ($ M) | Probability (%) |
|---|---|---|
| High | 6.3 | 20 |
| Medium | 6.3 | 50 |
| Low | 6.3 | 30 |
| Expected value ($ M) | 6.3 | |

(iii)    Flexible case

To incorporate flexibility into the project, we can analyse the consequence of building the facilities with the flexibility to add water injection at a later stage. Accordingly, we should make an additional investment of $30.0 M upfront, increasing the project's initial CAPEX from $710 M (in the base case or no injection) to $740 M.

The $30.0 M represents the first phase of the water injection facilities' investment that allows preparing the facilities for future construction of the water injection treatment and boosting if it is confirmed that the aquifer cannot support the oil production.

For completing the construction of the water injection facilities, an additional investment of $75.0 M is required.

According to the forecast, in the first two years, the three cases' production will be very similar; in the third year, the forecast will start differentiating between them depending on the aquifer's strength. In the flexible case, we can locate gauges in several wells to monitor the pressure closely; then, depending on the first indications of the pressure performance (aquifer support), we can decide whether the second phase of the water injection facilities' investment is required.

In this case, the three production profiles will be the same, as shown in Fig. 8.6.

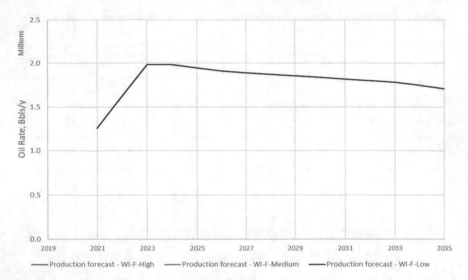

**Fig. 8.6** Production profiles with flexible water injection facilities

In the flexible case, the second stage of water injection facilities ($75.0 M) is made effective only if the production follows the production profiles medium and low; the investment to complete the facilities is the same in both cases. Figure 8.7 depicts these different cash flows.

Table 8.3 shows the NPVs and probabilities for each scenario and the expected value of this project.

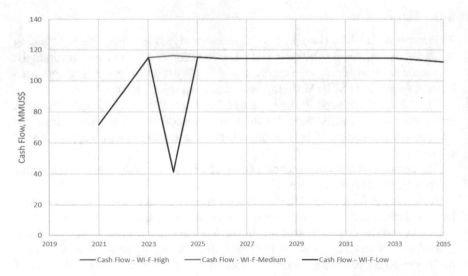

**Fig. 8.7** The cash flow of the project from the first oil for the flexible case

**Table 8.3** Financial assessment of the flexible water injection case

| Scenario | NPV ($ M) | Probability (%) |
|---|---|---|
| High | 76.3 | 20 |
| Medium | 25.1 | 50 |
| Low | 25.1 | 30 |
| Expected value ($ M) | 35.3 | |

As shown, the flexible case makes the expected value of the project reach $35.3 M, which is better than the base case ($−19.4 M) and the water injection case ($6.3 M). The flexible case significantly reduces the project's risk; all the flexible cases have a positive NPV, while the base case is extremely harmful ($−107.8 M). The water injection case has a moderate risk ($6.3 M).

In the project with flexibility, just US$30 MM should be invested upfront, and only in the case in which the aquifer is weak will an additional US$75 MM be invested in the year 2024. Two alternatives may happen: (i) the initial investment is made, and, after three years, it is confirmed that the water injection is not needed: in this case, the project's NPV is US$76.0 MM (in this case, the total investment is US$710 MM + 30 MM); and (ii) the initial investment is made, and, after three years, it is confirmed that the water injection is needed: in this case, the project's NPV is US$25.0 MM (in this case, the total investment is US$710 MM + 30 MM + 75 MM).

By adding flexibility to the project, the expectation is to increase the project value. Depending on the flexibility incorporated into the project, it can impact each case differently. In our example, the rationale of the flexibility is to reduce the downsize case because of the great exposure or risk to lose $107.8 M. The option implemented impact, at the same time the medium case, improving the value from losing $16.6 M to gaining $25.1. However, the flexibility added impact negatively the high case reducing the value from $106.3 to $76.3. The strategy implemented in this example focus on reducing the exposure to risk and improve the overall project value; this strategy is successfully implemented as shown above.

In other cases, the rationale of the flexibility is different: often, we are interested in improving the value of the high case regardless of the downsize case. In summary, flexibility should be implemented following the strategy that the decision-maker steer.

Looking back at this example, we can conclude that using the standard techniques for valuing the projects, the base case is not acceptable and carries a great risk. Even though the water injection case is a financially positive project, it just passes the minimum economic criteria, but it is not financially attractive. However, when flexibility is included, the project's value is improved significantly, and a highly positive project is delivered with a significant risk reduction.

The uncertainty in a project is often weighted through the discount rate in the NPV calculation: the higher the uncertainty, the larger the discount rate. This way of accounting for uncertainty makes logical sense only for portfolio evaluations when comparing projects and for ranking proposes; however, it has been used to assess single projects even though the justification is doubtful.

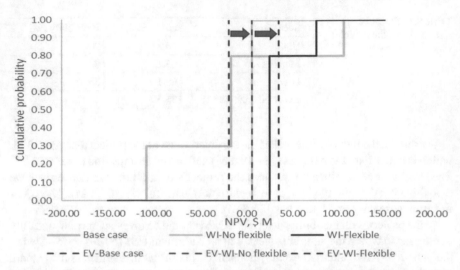

**Fig. 8.8** VARG diagram showing the impact of uncertainty reduction due to flexibility implementation

When flexibility is incorporated into the decision, because the most damaging cases are effectively removed from the evaluation, the NPV of the project is increased, and the VARG plot shown in Fig. 8.8 captures the example just discussed.

In Fig. 8.8, in yellow, we show the base case; in light green, the water injection case with no flexibility (full investment upfront) and red the water injection with flexibility. It can be observed that the low and medium cases have significantly shifted to the right; the low case moves from $-107.8 to a positive $25.1, which indeed reduce the project risk remarkably. The medium case also improves its value. The high case experiences a value reduction, but it is not as significant as the value improvement in the low case. The vertical dashed lines show the improvement in expected value comparing the base case (no water injection), the water injection without flexibility (water injection with investment upfront), and the water injection with flexibility. In summary, the expected values increase from a negative $-19.4 to a positive $35.3; the blue arrows highlight the estimated improvement. Implementing this project with flexibility can improve its value by avoiding the more pessimistic scenarios.

This example is an intuitive application of flexibility in a project valuation that clearly explains this approach's benefits. More sophisticated applications can be developed using Monte Carlo simulation, as we will show in Chap. 9 of this book. Several applications of the book's methodologies are explained in more detail.

## 8.7  Summary

The value of flexibility is one of the central topics of this book. We explain the rationale for creating flexibility in a project and how we can use it to create value in the project by taking advantage of project uncertainties. Creating flexibility using real options has several approaches which are discussed. Valuation methods are presented, and several types of options are described in detail. The value-at-risk-and-gain plot, which shows the increase in project value and risk reduction by using flexibility in the project, is explained. Several standard methods for value options are presented, and the difficulty of implementing them in engineering problems is acknowledged. Then, the engineering approach for valuing options is introduced, including their elements and workflow; a detailed application of the value of flexibility methodology in a case study in the oil and gas industry is presented.

## References

Amram, M., & Kulatilaka, N. (1999). *Real options: Managing strategic investments in an uncertain world*. Harvard Business School Press.

Babajide, A., de Neufville, R., & Cardin, M. (2009). Integrated method for designing valuable flexibility in oil development projects. *Society of Petroleum Engineers*. In *Proceeding of the 2009 SPE Projects*, Facilities & Construction. Society of Petroleum Engineers. Paper SPE-122710-PA.

Begg, S., & Bratvold, R. (2002). The value of flexibility in managing uncertainty in oil and gas investments. *Society of Petroleum Engineers*. In *Proceeding of the SPE Annual Technical Conference and Exhibition*, San Antonio, Texas, USA, 29 September–2 October 2002. Paper SPE-77586.

Begg, S., Bratvold, R., & Campbell, J. (2004). Abandonment decisions and value of flexibility. *Society of Petroleum Engineers*. In *Proceeding of the SPE Annual Technical Conference and Exhibition*, Houston, Texas, USA, 26–29 September 2004. Paper SPE-91131.

Binder, W., Paredis, C., & Garcia, H. (2017, December). The value of flexibility in the design of hybrid energy systems: A real options analysis. *IEEE Power and Energy Technology System Journal, 4*(4), 74–83.

Brandao, L., Dyer, J., & Hahn, W. (2005, June). Using binomial decision trees to solve real-option valuation problems. *Decision Analysis, 2*(2), 69–88.

Coopersmith, E., & Cunningham, P. (2002). A practical approach to evaluating the value of information and real options decision in the upstream petroleum industry. *Society of Petroleum Engineers*. In *Proceeding of the SPE Annual Technical Conference and Exhibition*, San Antonio, Texas, USA, 29 September–2 October 2002. Paper SPE-77582.

Copeland, T., & Antikarov, V. (2001). *Real options, A practitioner's guide* (1st ed.). Texere LLC.

Cox, J., Ross, S., & Rubinstein, M. (1979, October). (1979) Option pricing: A simplified approach. *Journal of Financial Economics, 7*, 229–264.

Dezen, F., & Morooka, C. (2002). Real options applied to the selection of technological alternative for offshore oilfield development. *Society of Petroleum Engineers*. In *Proceeding of the SPE Annual Technical Conference and Exhibition*, San Antonio, Texas, USA, 29 September–2 October 2002. Paper SPE-77587.

Han, J. (2003). There is a value in operational flexibility: An intelligent well application. *Society of Petroleum Engineers*. In *Proceeding of the SPE Hydrocarbon Economics and Evaluation Symposium*, Dallas, Texas, USA, 5–8 April 2003. Paper SPE-82018.

Jafarizadeh, B., & Bratvold, R. (2009). Real options analysis in petroleum exploration and production: A new paradigm in investment analysis. *Society of Petroleum Engineers.* In *Proceeding of the 2009 SPE EUROPEC/EAGE Annual Conference and Exhibition*, Amsterdam, The Netherlands, 8–11 June 2009. Paper SPE-121426.

Kim, K., Kim, S., & Kim, H. (2016). Real options analysis for photovoltaic project under climate uncertainty. In *Proceedings of the 2016 International Conference on New Energy and Future Energy System (NEFES 2016). IOP Conference Series: Earth and Environmental Science, 40*(2016), 012080. https://doi.org/10.1088/1755-1315/40/1/012080.

Muller, W. (2018). *Simulated real options approach to investment decisions of Dutch dairy farmers.* M.Sc. thesis, Wageningen University, Business Economics Group, Business Economics–Management, Economics and Consumer Studies.

Myers, S. (1984, Jan–Feb). Finance theory and financial strategy. *Interfaces, 14*, 126–137.

Neufville, R., Scholtes, S., & Wang, T. (2006). Real options by spreadsheet: Parking garage case example. *American Society of Civil Engineers. Journal of Infrastructure Systems, 12*(3), 107–111.

Trigeorgis, L., & Mason, S. (1987, Spring). Valuing managerial flexibility. *Midland Corporate Finance Journal, 5*(1), 14–21.

Wang, H. (2007, March). *Sequential optimisation through adaptive design of experiments* (Doctoral thesis). Massachusetts Institute of Technology, Engineering Systems Division.

Willigers, B., & Bratvold, R. (2008). Valuing oil and gas options by least square Monte Carlo simulation. *Society of Petroleum Engineers.* In *Proceeding of the 2008 SPE Annual Technical Conference and Exhibition*, Denver, Colorado, USA, 21–24 September 2008. Paper SPE-116026.

# Chapter 9
# Case Studies for the Value of Information and Flexibility in the Oil and Gas Industry

**Objective**

Four case studies are used to show the methodologies discussed in this book. The first case study shows a complete value of information assessment using the decision maker's risk attitude. The second case study shows an example of the value of flexibility in an oil development example. The third case study uses the design of experiments methodology to optimize the value of information assessment. Finally, the fourth case study incorporates the data's fuzziness to the value of information discussed in the third case study.

## 9.1  Examples of the Value of Information

Alpha field, located in the north of Africa, has been explored and successfully discovered. One exploration well and three appraisal wells were drilled, confirming the potentiality of this field. The area of the reservoir allows for preparing a field development plan. Two of the exploratory wells reached the oil-to-water contact. The PVT sample taken in one of the wells suggests that the oil has low gas saturation, meaning that gas expansion is not expected to contribute to the oil production.

There is no evidence to suggest the level of support that the aquifer will provide to oil production. Based on the performance of analogous fields, three scenarios have been prepared for the development of the field, assuming strong, medium, and weak support from the aquifer; these scenarios assume five wells to hook up in the first year and another five wells in the second year. These production scenarios are shown in Fig. 9.1.

The expert team members working on the asset estimate that the aquifer's chances to provide strong, medium, and weak support are 40, 35, and 25%. The oil price is assumed as \$50/Bbls during the year 2022 with a 5% annual increase. The discount rate used in the financial evaluation is 10%.

© The Author(s), under exclusive license to Springer Nature Switzerland AG 2022       251
M. J. Vilela and G. F. Oluyemi, *Value of Information and Flexibility*,
Petroleum Engineering, https://doi.org/10.1007/978-3-030-86989-2_9

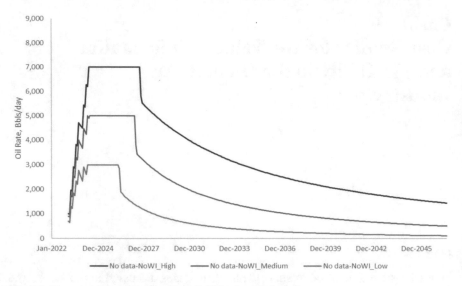

**Fig. 9.1** Forecasted scenarios for Alpha field for natural depletion

The surface facility cost to handle oil production and deliver it to the main facility is estimated at \$350 M. Well-drilling, logging, stimulation, completions, and flowlines to the facility are estimated at \$10 M.

The CAPEX of the project was estimated to be \$450 M, including the cost of the ten wells (\$10 M each). Based on nearby fields, the OPEX was estimated at \$2.0/Bbls.

An exponential utility function is used to account for the risk attitude of the company; based on the previous project, the risk tolerance (RT) factor used is \$200 M. Figure 9.2 shows the utility function used in this assessment.

Table 9.1 shows a summary of the main financial parameters.

The expected value of this project is \$61.3 M.

Figure 9.3 shows the decision tree associated with the value and the utility value of this project.

The expected value of the utility is zero, as shown in Fig. 9.4.

Even though the project's expected value is positive (\$61.3 M), there is a 25% chance that the project will result in a loss of \$238.8 M. This means that there is a significant possibility of losing a relevant amount of money. This risk can be assessed differently for different companies, which is captured in the utility function. Figure 9.3 indicates that it is better not to move forward with the project but to cancel it.

After assessing the estimated expected gain and the risk associated, the company's management was reluctant to move forward with the project under the conditions outlined. The technical team indicated that the main reason for the mediocre financial evaluation is the small amount of oil that can be recovered if the aquifer provides medium or poor production support. The technical solution for this issue is to inject water into a few wells for water sweeping and reservoir pressure maintenance.

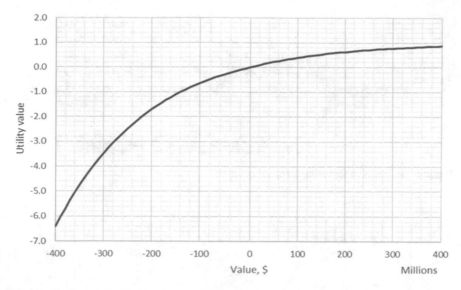

**Fig. 9.2** Exponential utility function used in the case study

**Table 9.1** Summary of financial parameters of the original project

|  | Chance of occurring, % | Cumulative oil, M Bbls | NPV value, M $ | Utility value |
|---|---|---|---|---|
| High case | 40 | 31.2 | 297.9 | 0.7745 |
| Mid case | 35 | 16.8 | 5.2 | 0.0257 |
| Low case | 25 | 6.8 | −238.8 | −2.2995 |

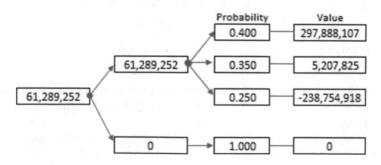

**Fig. 9.3** Decision tree for the expected value

An evaluation of this project subject to water injection shows that a significant improvement in oil production is observed when water is injected, as shown in Fig. 9.5.

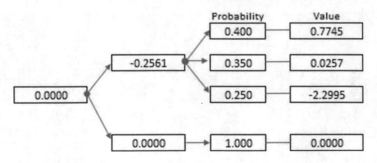

**Fig. 9.4**  Decision tree for the expected utility value

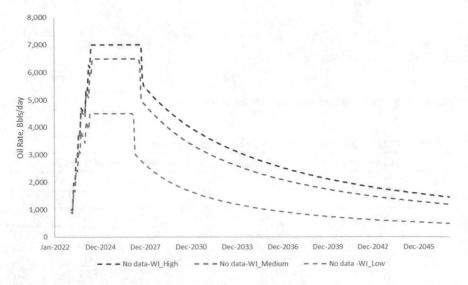

**Fig. 9.5**  Forecasted scenarios for Alpha field for water injection

The medium and low cases with water injection show an improved recovery compared with the natural depletion cases. This statement is supported by Fig. 9.6, which compares the cumulative oil recovery.

It is observed that with water injection, the high case remains unchanged, and the medium and low cases improve by 61% and 118%, respectively.

The project, including water injection upfront, shows better financial performance than the natural depletion scenario. Figure 9.7 shows the expected value of the upfront water injection project.

The same assessment using utility values is shown in Fig. 9.8.

The comparison of the evaluations shown in Figs. 9.3 and 9.7 indicate an increase in monetary value from $61.3 M to $90.7 M, i.e., $29.4 M.

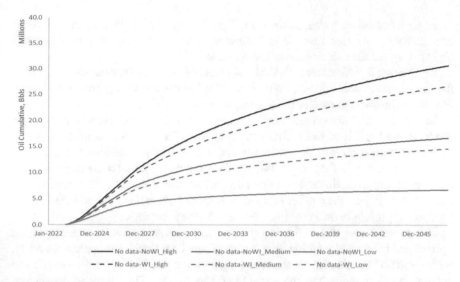

**Fig. 9.6** Forecasted cumulative oil scenarios for Alpha field for natural depletion and water injection

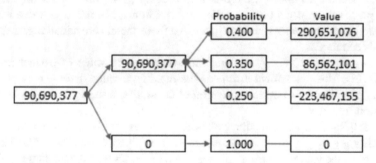

**Fig. 9.7** Decision tree for monetary value on water injection

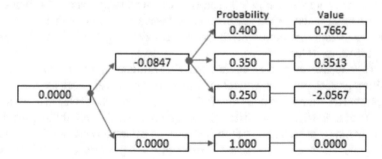

**Fig. 9.8** Decision tree for utility value on water injection

In terms of utility values, shown in Figs. 9.4 and 9.8, both scenarios have an expected value of zero because in both cases, the move forward scenario is still highly risky for the risk attitude of the company.

In the water injection scenario, CAPEX was adjusted to include the water injection facility; several water injector wells were added to optimise oil production: 10 and 5 wells in the low and medium cases, respectively.

Expert team members suggest a data acquisition alternative. In this case, only one producer well is drilled in the first year, and it produces for six months; after that period, an evaluation is made whether the aquifer is strong, medium or weak.

This scenario with data acquisition delays the development for six months while the data is acquired. However, in this case, a better assessment of the main uncertainty of the project, the aquifer support, is acquired. An adjustment in CAPEX and production profiles is made for this new development scenario.

Figure 9.9 shows the production profiles of the scenarios with water injection upfront and the data acquisition. In terms of production, they are remarkably similar, with a shift of six months for the data acquisition case; this is the time needed to acquire the data and assess the strength of the aquifer. This delay in production causes some losses in revenues during the first year, impacting negatively on the project's economy. However, there is also a delay in the investment for facilities, which positively impacts the project's financial figures. In addition, there is a cost associated with the data acquisition itself. All these factors are added together in the project's economics.

As explained in Chap. 7, the first assessment is the value of perfect information (VOPI). This assessment estimates the maximum value that the data provides, assuming that it gives a perfect indication of the aquifer's strength. Figure 9.9 shows a decision tree in VOPI.

Figure 9.9 shows that if the first well is able to assess, without uncertainty, whether the aquifer is strong, medium or weak, the value of that information is $58.7 M; this result makes the value of perfect information 65% better than the no data acquisition (water injection upfront). This assessment indicates that the best option is acquiring the data and developing the field based on that information. However, that case is unrealistic because data is not perfect, and it always has uncertainties.

For considering the company attitude towards risk, Fig. 9.10 shows the decision tree for the utility values.

Figure 9.10 shows that, including the risk attitude of the company, the data acquisition project assuming perfect information is positively evaluated.

Experts in the team assess the "reliability" of the data, i.e., the chances that the data accurately describe the several states of the aquifer. Once the "reliability probability" is estimated, the "posterior probabilities" can be calculated using Bayes' theorem (see Chap. 7). Table 9.2 summarises the reliability and posterior probabilities, where a small letter indicates the data "prediction", h = high aquifer, and so on. The capital letter indicates the actual state, M = medium aquifer, and so on.

The value of information assuming the data is imperfect is shown through the decision tree in Fig. 9.11.

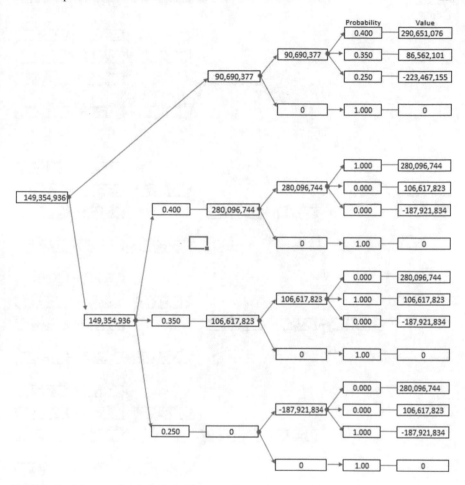

**Fig. 9.9** Decision tree for the value of data acquisition for water injection alternative with perfect information

The value of the data when the data is imperfect is $24.5 M. This result represents a drop in value from $58.7 M to $24.5 M from the case of perfect to imperfect information. However, the value of the project with data acquisition is higher than the value of the project without data acquisition. This assessment concludes that, in this case, study, the optimum option is to acquire the data and, based on those results, move forward with the project.

A similar analysis using utility values allows the risk attitude of the company to be captured. Figure 9.12 shows the decision tree for the utility values.

Figure 9.12 shows that considering the risk attitude of the company, this project with data acquisition is positively evaluated, suggesting moving forward to acquire the data and, based on those results, decide the fate of the project. In this assessment, both the expected value and the expected utility lead to the same recommended action.

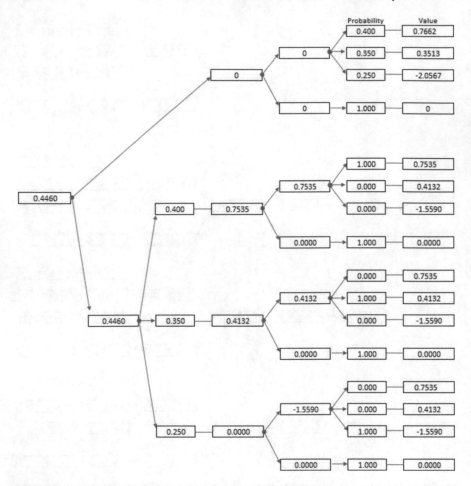

**Fig. 9.10** Decision tree for the utility value of data acquisition for water injection alternative with perfect information

**Table 9.2**  Reliability and posterior probabilities

| *Reliability probabilities* | | | | | | | | |
|---|---|---|---|---|---|---|---|---|
| P(h\|H) | P(m\|H) | P(l\|H) | P(h\|M) | P(m\|M) | P(l\|M) | P(h\|L) | P(m\|L) | P(l\|L) |
| 0.65 | 0.30 | 0.05 | 0.20 | 0.60 | 0.20 | 0.10 | 0.35 | 0.55 |
| *Posterior probabilities* | | | | | | | | |
| P(H\|h) | P(H\|m) | P(H\|l) | P(M\|h) | P(M\|m) | P(M\|l) | P(L\|h) | P(L\|m) | P(L\|l) |
| 0.73 | 0.29 | 0.09 | 0.20 | 0.50 | 0.31 | 0.07 | 0.21 | 0.60 |

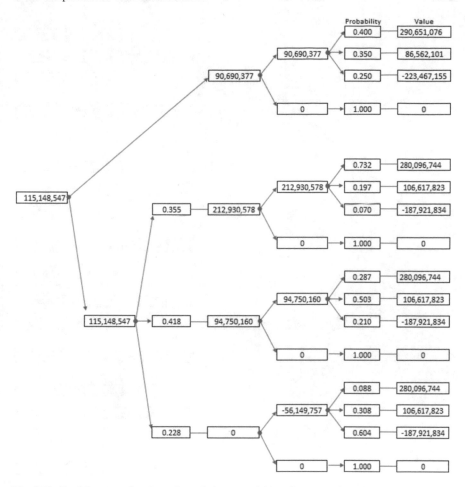

**Fig. 9.11** Decision tree for the value of data acquisition for water injection alternative with imperfect information

In summary, the original project assuming natural depletion, even though it has a positive value, carries many risks, and the utility value assessment suggests not continuing to the development phase.

Inject water in the reservoir could be implemented, from the beginning of production, for improving the oil production and, hopefully, the value of the project. This case enhances the value of the project but still has low utility due to the risk associated.

The third analysis is assumed to acquire data to reduce the uncertainties and improve the project's value. This scenario improves both the value and the project's utility, making this option the optimum to develop the field.

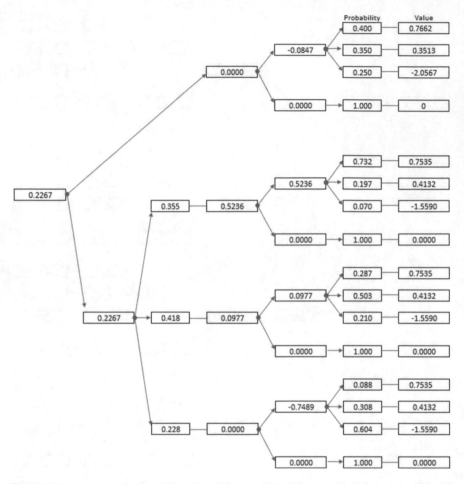

**Fig. 9.12** Decision tree for the utility value of data acquisition for water injection alternative with imperfect information

## 9.2   Examples of the Value of Flexibility

The Beta field is an offshore exploration development. For contractual commitments, Beta should start production next year.

Based on the production test carried out in the appraisal wells, reservoir compartmentalisation can strongly impact the field development. The seismic resolution did not identify faults or any other feature responsible for possible compartmentalisation. However, this can be the consequence of faults below the seismic resolution or diagenetic phenomena.

Assuming an Original Oil In Place of 200 M Bbls, it was estimated that sixteen wells could be located in the development plan. Three scenarios with sixteen wells each are built for developing the field under three levels of compartmentalisation:

the high production scenario that corresponds to the low compartmentalisation level, the medium production scenario associated with the mid-level of compartmentalisation, and the low production scenario that corresponds to the high level of compartmentalisation.

Figure 9.13 shows the production scenarios for the three cases.

These three scenarios generate the cumulative production shown in Fig. 9.14.

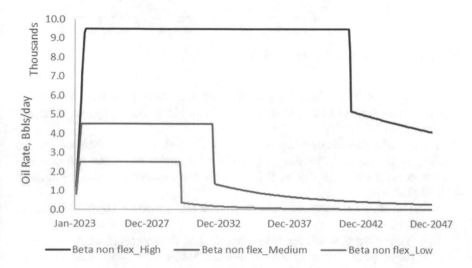

**Fig. 9.13** The production forecast for the non-flexible scenarios of Beta

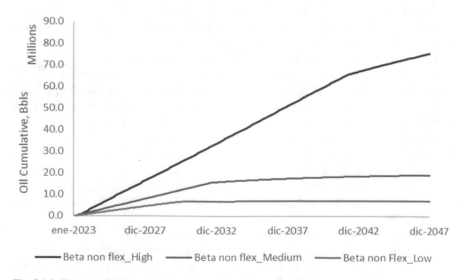

**Fig. 9.14** The cumulative production forecast for the non-flexible scenarios of Beta

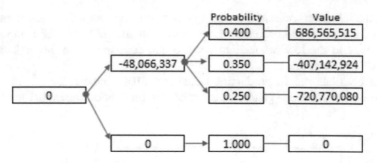

**Fig. 9.15** Decision tree for the non-flexible design of Beta

The expert team members assessed the chance of those scenarios representing the future production of Beta, assigning 40, 35 and 25% to the high, mid and low scenarios.

These production forecasts assume the construction of an oil platform, process facility, wells umbilical, flowline, engineering and management cost and slots and drilling cost for ten wells. The total CAPEX for this non-flexible case is estimated at $960.0 M.

Based on other fields developed under similar infrastructure, OPEX is estimated at $3.0/Bbls.

Due to the current world financial situation and the estimated world economic growth, the oil price is forecast at $50.0/Bbls in 2022 with a 2% increase per year.

The net present value (NPV) is used to measure a project's profitability (value). The discount factor used is 10%.

This design is called non-flexible because it assumes that a decision is taken upfront to develop the field with one configuration (facilities and wells). No change is permitted during the production life of the project.

Figure 9.15 shows the decision tree for this non-flexible design.

This assessment indicates that no future gain is expected from this project for the investment of $960.0 M. In this scenario, the best option is to abandon the project.

This project's design includes the uncertainty in the forecast due to the main uncertainty of this project, which is the reservoir's compartmentalisation. However, this design lacks any flexibility that can be incorporated into the design.

One possible area of flexibility is to drill just ten wells initially and build the platform and facilities for further expansion, depending on the first five years of production. After that period, a decision can be taken to drill the remaining six wells or some of them, depending on the observations captured during the first five years of production.

Figure 9.16 shows the decision tree for the flexible scenario.

Comparing the flexible case with the non-flexible case for the Beta field, an increase in the project's value of $51.4 M is observed by adding flexibility to the design. Including flexibility transforms a non-financially viable project into a positive project.

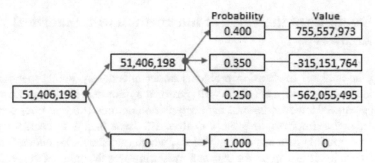

Fig. 9.16  Decision tree for the flexible design of Beta

As mentioned in Chap. 8, the Value-At-Risk-and-Gain (VARG) plot is frequently used in the financial community to represent the range of uncertainties of the project in value terms. In this example, VARG can show the changes in uncertainty due to changes in development strategies. Figure 9.17 shows the VARG for the non-flexible and flexible cases and the expected values for both cases.

VARG shows that the downside case has been reduced by incorporating flexibility in the design. Also, additional opportunities have been captured, as observed from the increase in the value for the high case.

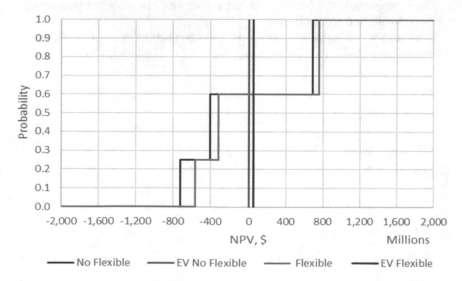

Fig. 9.17  VARG plot for non-flexible and flexible cases

## 9.3    Examples of the Value of Information with Statistical Analysis

During the analysis of decision problems under uncertainty, sometimes one data acquisition action (take a core in a well, perform a pilot test, etc.) can improve the project's value. However, several data acquisition actions (no just one) can often be considered to improve the project's value. The issue then is to decide, amongst the several possible data acquisition actions, which one should be undertaken. The rational decision is to acquire the data that most improves the value of the project. To do that, experimental design (DOE), described in Chap. 5, is the optimum technique.

This case study refers to an oil field development where the three main uncertainties are: (i) the average reservoir permeability, (ii) the aquifer support to production, and (iii) the size of the reservoir. This latest uncertainty occurs because the seismic image shows both the main structure and potentially a small structure near to it. Exploration wells confirm that the main structure is prospective, but no wells have yet explored the small structure.

Based on what we learn in Chap. 5, for screening between the three variables and determining which ones are the most important, we should conduct eight experiments using high (+) and low (−) values for each variable.

Figure 9.18 shows the production profiles for the eight experiments.

The three symbols in brackets refer to the high and low values of permeability, aquifer strength, and reservoir size.

In all cases, we keep a plateau for 1.0–1.5 years. Figure 9.19 shows the corresponding cumulative oil.

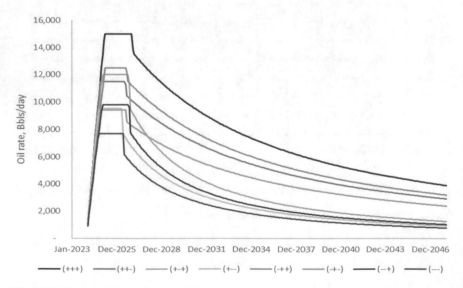

**Fig. 9.18**  Screening for the three variables

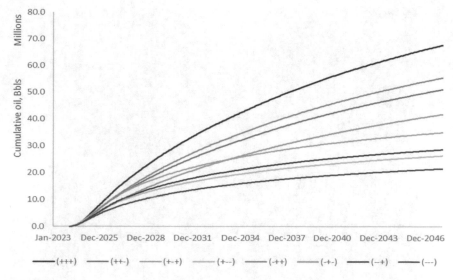

**Fig. 9.19**  Cumulative oil for the screening

Each experiment is evaluated using the net present value. CAPEX is $1.040 M if the development is only in the main structure and $1.220 M for both structures. The oil price is assumed at $50/Bbls in 2023, with a 2% increase per year. The project lasts for 25 years.

The discount rate used is 10%. OPEX is $3.0/Bbls. For developing the main structure, twelve wells are drilled, and for the small structure, four wells.

Table 9.3 shows the NPV of the eight runs of this experiment.

Figure 9.20 shows a 3D representation of the experimental design, the cube plot.

Statistical analysis of these values indicates that the main parameter (the one whose variation impacts the value of the project most) is the aquifer's strength, followed by the permeability and, finally, the size of the reservoir. This conclusion can be reached by observing the Pareto plot of the experiment, shown in Fig. 9.21.

**Table 9.3** NPV for the eight runs of the screening experiment

| Experiment | NPV, $ |
|---|---|
| (+++) | 496,493,364 |
| (++−) | 252,156,914 |
| (+−+) | −208,441,699 |
| (+−−) | −283,883,022 |
| (−++) | 187,584,697 |
| (−+−) | 17,210,770 |
| (−−+) | −389,396,548 |
| (−−−) | −421,443,378 |

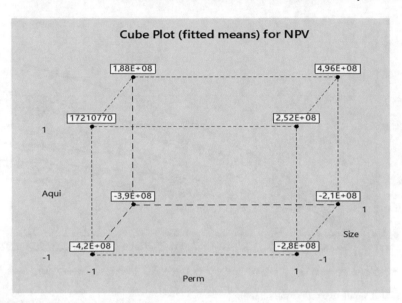

**Fig. 9.20** Cube plot for the experiment

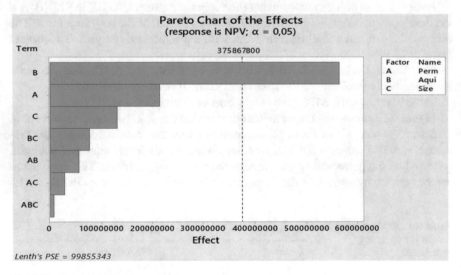

**Fig. 9.21** Pareto plot for the experiment

This experiment is assessed using a 5% significance level (see Chap. 3). A similar conclusion is reached using the Normal plot, shown in Fig. 9.22.

The effect of the parameters is shown in Fig. 9.23, where it is observed that a change in factor B, the aquifer strength, produces the most significant change in the NPV; factor A is the next in relevance, but its impact is far from factor B.

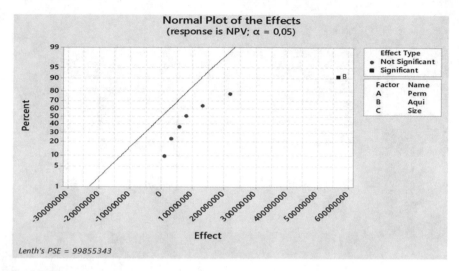

**Fig. 9.22** Normal plot for the experiment

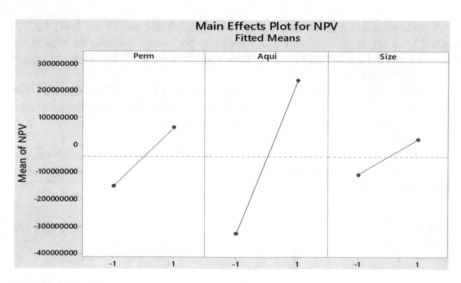

**Fig. 9.23** Main effects

Figure 9.23 shows that the three effects increase when the parameter increases from −1 to +1. However, factor B is the most relevant, followed by factor A and then factor C.

It is essential to assess interaction effects because they can produce a significant impact on the value. Sometimes, the main effects are the result of a hidden interaction effect. Figure 9.24 shows the interaction effects.

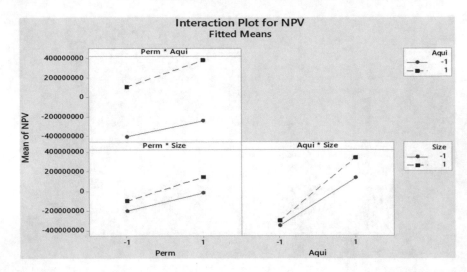

**Fig. 9.24** Interaction effects

For the interaction effects in parallel, there is no difference for the second effect value; however, in the case of aquifer support, the interaction with the size effect suggests that the greater the size of the reservoir, the greater will be the enhancement due to aquifer support.

Another display that helps to better understand the impact of the factors is the surface plot. Figure 9.25 shows the surface plot for the aquifer and permeability, where we can see that the optimum solution is high aquifer support and high permeability.

These results indicate that the uncertainty that most impacts the project's value is the aquifer strength (factor B). In contrast, the second factor is permeability (factor A), and the third factor is reservoir size (factor C).

**Fig. 9.25** Surface plot for aquifer strength and permeability

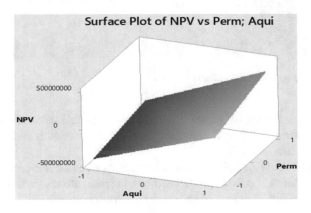

In addition, aquifer strength is the only statistically significant factor; this conclusion means that the variations observed in the outcomes result from the uncertainty in this factor.

The importance of this analysis is that it indicates the important uncertainties, in terms of the project's value, that are worth focusing on and those which are not relevant. This determination drives the data acquisition actions that the decision-maker should follow.

In our case study for each uncertainty, we can identify a data acquisition action that can be taken to reduce the uncertainty in the variable. For the uncertainty in permeability, a core can be taken in a new well and a conventional core analysis performed; of course, only lab analysis is required if a core already exists. In the case of uncertainty in the aquifer strength, a well can be produced for a few months, and the well performance can be analysed to conclude if the aquifer provides support and assesses that support. For this temporary production, fluid production can be treated in low capacity temporary/portable facilities or transported to nearby facilities. Whether a small extension in the field is present or not, an additional well can be drilled in that extension area.

Based on the statistical analysis of the data provided in Table 9.3, the only uncertainty of significance to the project's value is the aquifer strength. Thus, the data acquisition action suitable for this case is anything that can determine the aquifer's strength.

The other two data acquisition actions, the core analysis and the additional well in the extension area, even though they can reduce the uncertainty in these two variables, are not worthwhile because the change in the value of the project resulting from them falls below the 5% significance level in the normal variability in the data.

The aquifer strength can be defined by drilling a well and produce it for six months (data acquisition action); depending on the well performance, an indication of the aquifer strength can be derived. For this short production period with one well, the fluid will be treated in a temporary facility. The deployment of this facility has an associated cost. Additionally, the project will carry a six-month delay compared with the no data acquisition project.

Figure 9.26 shows the oil rate and cumulative oil for the high, medium and low scenarios assuming no data acquisition.

If a data acquisition action is taken, oil production is from one well only for six months, and the full development will start later. Figure 9.27 shows the respective profiles.

The oil production and the CAPEX are delayed for six months in the data acquisition case. The additional cost for data acquisition is estimated at $5.0 M. Table 9.4 shows the NPV for the three scenarios for the no data acquisition and the data acquisition cases.

The team of experts assessed the probability of occurrence of the no data acquisition scenarios. Furthermore, they evaluated the reliability probability of the test (evaluation of aquifer strength based on first well performance) and computed the posterior probabilities using Bayes' theorem. The probabilities are presented in Table 9.5.

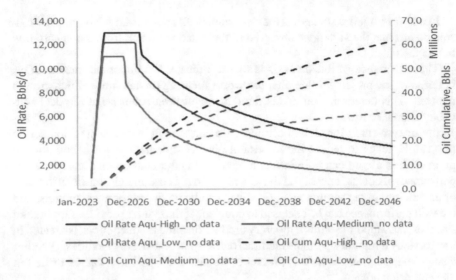

**Fig. 9.26** Oil forecast for high, medium and low scenarios for no data acquisition scenarios

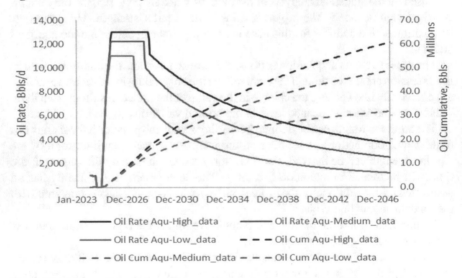

**Fig. 9.27** Oil forecast for high, medium and low scenarios for data acquisition scenarios

| **Table 9.4** NPV for the no data acquisition and data acquisition scenarios | Scenario | NPV, $ M |
|---|---|---|
| | Aqu-High_No Data | 340.2 |
| | Aqu-Mid_No Data | 92.2 |
| | Aqu-Low_No Data | −298.9 |
| | Aqu-High_Data | 311.4 |
| | Aqu-Mid_Data | 78.3 |
| | Aqu-Low_Data | −285.2 |

**Table 9.5** Reliability and posterior probabilities in the case study 9.3

| Probability with no data acquisition | | Reliability probabilities | | Posterior probabilities | |
|---|---|---|---|---|---|
| | | Prob(h|H) | 0.700 | Prob(H|h) | 0.770 |
| Prob(H) | 0.320 | Prob(m|H) | 0.200 | Prob(M|h) | 0.170 |
| Prob(M) | 0.330 | Prob(l|H) | 0.100 | Prob(L|h) | 0.060 |
| Prob(L) | 0.350 | Prob(h|M) | 0.150 | Prob(H|m) | 0.184 |
| Probability with data acquisition | | Prob(m|M) | 0.650 | Prob(M|m) | 0.615 |
| | | Prob(l|M) | 0.200 | Prob(L|m) | 0.201 |
| Prob(h) | 0.291 | Prob(h|L) | 0.050 | Prob(H|l) | 0.089 |
| Prob(m) | 0.349 | Prob(m|L) | 0.200 | Prob(M|l) | 0.183 |
| Prob(l) | 0.361 | Prob(l|L) | 0.750 | Prob(L|l) | 0.728 |

Capital letters are used for the *state of nature*, high, medium and low. Small letters are used for the data assigned to the states of nature. Prob(H) is the probability of occurrence given to the *state of nature* "high". Prob(l) is the probability that the data predict that the state "low" will occur. Prob(m|H) is the probability that the *state of nature* is "high", and the data assign the state as "medium". Probl(M|l) is the probability that, given the data, assign the state as "low", while the *state of nature* is at the "medium" level.

Figure 9.28 shows the decision tree for the no data acquisition.

This project has an NPV of $34.7 M.

Figure 9.29 shows the decision tree for the data acquisition proposed in this case study.

The NPV is $85.4 M for the data acquisition scenario, an increment of $58.7 M, more than double the value of the project without data acquisition. In this case study, the result analysis encourages data acquisition before engaging in the project.

This case study strongly supports the use of DOE to rank the possible data acquisition actions associated with a project and implement only those that improve the value the most.

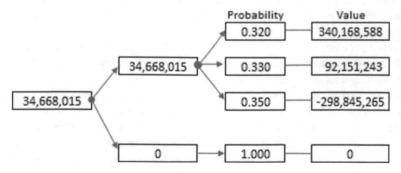

**Fig. 9.28**  Decision tree for the no data acquisition scenario

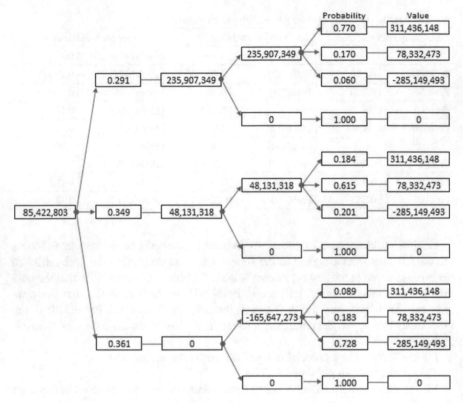

**Fig. 9.29** Decision tree for the data acquisition scenario

## 9.4   Fuzzy Data Acquisition

As mentioned in Chap. 6, probability captures the uncertainty of a result. For data acquisition, we estimate the probability of the several scenarios and the reliability probability, i.e., the probability that data acquisition results in one value given that the *state of nature* is given. These calculations are justified on the uncertainty of the *state of nature* and data outcomes.

There is another form of uncertainty related to the precision of the data and the fuzziness associated with the data. This kind of uncertainty is captured by using fuzzy logic and membership functions.

In the case study discussed in Sect. 9.3, we assume that the aquifer will provide an answer that indicates which type of support the aquifer will provide: strong, medium, or weak, without considering ambiguity, either one or the other. However, there is some level of ambiguity in that assessment, and it can be captured using fuzzy logic.

This case study will assume that three fuzzy sets describe aquifer support: high support, medium support, and low support. These sets are called: $\tilde{M}_h$, $\tilde{M}_m$ and $\tilde{M}_l$. The membership functions represented by these sets capture the fuzziness of each of

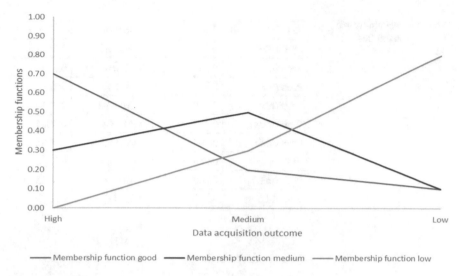

**Fig. 9.30** Membership functions for the case study 9.4

the outcomes. For example, if the outcome of the test suggests high support, we are weighting it as 0.7, but 0.2 for medium and 0.1 for low; these figures indicate that, due to ambiguity in the measurements or the analysis of the results, the level "high" represents high support (with a weight 0.7 out of 1.0), but also it represents medium support (0.2 out of 1.0) and low support (0.1 out of 1.0).

The membership functions are built by the expert team members assessing the uncertainties due to the fuzziness in the aquifer support. The complete set of membership functions designed for this case study is shown in Fig. 9.30.

By using Eqs. (7.23) and (7.24), the value of information can be transformed from a crisp problem to a fuzzy problem through the membership function. The evaluation of the value of the problem uses the expected value of the probabilities of each *state of nature* depending on the corresponding probability of the fuzzy sets. Equations (7.26)–(7.29) are the critical equations required to perform the value of information analysis with fuzzy data.

Table 9.6 shows the reliability probabilities and the posterior "fuzzy" probabilities calculated using the equations referred to above.

The uncertainty due to data ambiguity can be described using a fuzzy set, and the value of the project is estimated, as shown in Fig. 9.31.

In the comparison between the value of the crisp data with the fuzzy data, it is observed that the value of the data reduces when the uncertainty due to fuzziness is incorporated in the analysis. The value of the project with crisp data is $85.4 M (Fig. 9.29) compared with $68.5 M for the fuzzy data. This result is expected as another layer of uncertainty is added to the problem.

Two sensitivities are performed; it can be shown that when the membership functions are changed in the manner shown in Fig. 9.32, the fuzziness in the data is removed.

**Table 9.6** Reliability and "fuzzy" probabilities

| Reliability probability | | Posterior "fuzzy" probability | |
|---|---|---|---|
| Prob (h\|H) | 0.700 | Prob (H\|$\tilde{M}_H$) | 0.5584 |
| Prob (m\|H) | 0.200 | Prob (H\|$\tilde{M}_M$) | 0.3441 |
| Prob (l\|H) | 0.100 | Prob (H\|$\tilde{M}_L$) | 0.1140 |
| Prob (h\|M) | 0.150 | Prob (M\|$\tilde{M}_H$) | 0.2719 |
| Prob (m\|M) | 0.650 | Prob (M\|$\tilde{M}_M$) | 0.4325 |
| Prob (l\|M) | 0.200 | Prob (M\|$\tilde{M}_L$) | 0.2981 |
| Prob (h\|L) | 0.050 | Prob (L\|$\tilde{M}_H$) | 0.1697 |
| Prob (m\|L) | 0.200 | Prob (L\|$\tilde{M}_M$) | 0.2235 |
| Prob (l\|L) | 0.750 | Prob (L\|$\tilde{M}_L$) | 0.5879 |

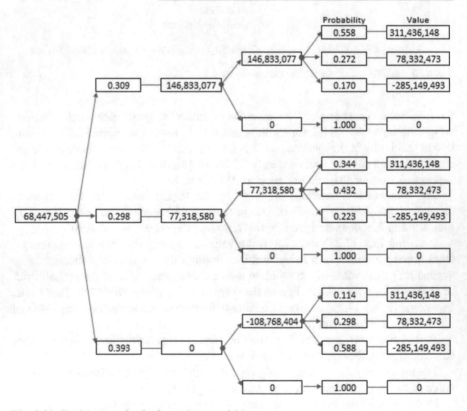

**Fig. 9.31** Decision tree for the fuzzy data acquisition

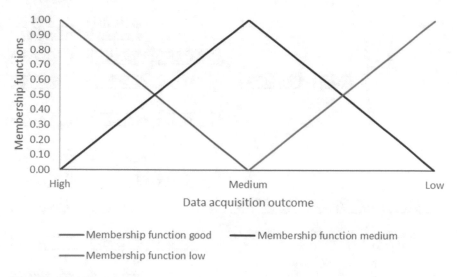

**Fig. 9.32** Sensitivity on the membership functions to remove the fuzziness

The value of the project is again $85.4 M because those membership functions remove the uncertainty due to ambiguity. Also, if the membership functions are modified to increase the fuzziness, such as in Fig. 9.33.

The value of the project reduces even further to $34.2 M, which is lower than the value of the no data acquisition case. Figure 9.34 shows the decision tree for the fuzzy sets shown in Fig. 9.33.

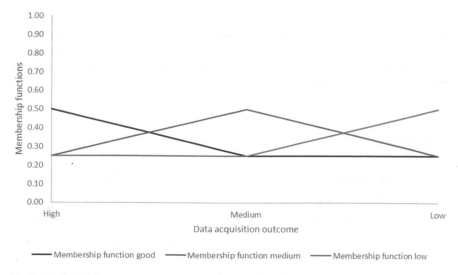

**Fig. 9.33** Sensitivity on the membership functions to increase the fuzziness

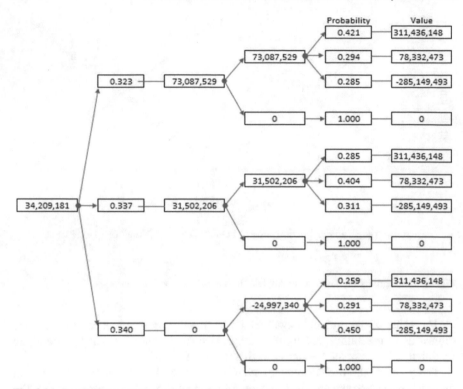

**Fig. 9.34** Sensitivity on the membership functions for the fuzzy data acquisition to increase the fuzziness

The determination of the accurate membership function for each problem is an open problem. With this example, we show the impact that the uncertainty due to fuzziness can have in evaluating a decision problem and how this uncertainty can be addressed. Further investigation is required for the definition of the most accurate fuzzy set for each problem.

## References

Abadie, L., & Chamorro, J. (2017). Valuation of real options in crude oil production. *Energies, 10*(1218), 1–21.

Abdul Aziz, P., Ariadji, T., Fitra, U., & Grion, N. (2017). The implementation of real option theory for economic evaluation in oil and gas field project: Case studies in Indonesia. *International Journal of Applied Engineering Research, 12*(24), 15759–15771.

Allais, M. (1953, October). Le Comportement de l'Homme Rationnel devant le Risque: Critique des Postulats et Axiomes de l'Ecole Americaine. *Econometrica, 21*(4), 503–546.

Arild, O., Lohne, H., & Bratvold, R. (2008). A Monte Carlo approach to value of information evaluation. *Society of Petroleum Engineers*. In *Proceeding of the International Petroleum Technology Conference*, Kuala Lumpur, Malaysia, 3–5 December. Paper IPTC-11969.

Banerjee, A., Chassang, S., & Snowberg, E. (2016). *Decision theoretic approaches to experimental design and external validity.* National Bureau of Economic Research, Working paper 22167, Cambridge, Massachusetts, USA.

Bardakhchyan, V. (2017). Fuzzy Bayesian inferences. *Physical and Mathematical Sciences, 51*(1), 8–12.

Barton, R. (2013). Designing simulation experiments. In R. Pasupathy, S.-H. Kim, A. Tolk, R. Hill, & M. E. Kuhl (Eds.), *Proceedings of the 2013 Winter Simulation Conference* (pp. 342–353).

Begg, S., Bratvold, R., & Welsh, M. (2014). Uncertainty vs. variability: What's the difference and why is it important? *Society of Petroleum Engineers.* In *Proceeding of the SPE Hydrocarbon Economics and Evaluation Symposium,* 19–20 May, Houston, Texas, USA. Paper SPE-169850.

Berger, P., Maurer, R., & Celli, G. (2018). *Experimental design with applications in management, engineering, and the sciences* (2nd ed.). Springer.

Bezdek, J. (1993, February). Fuzzy models—What are they, and why? *IEEE Transactions on Fuzzy Systems, 1*(1), 1–6.

Binmore, K. (2017). On the foundations of decision theory. *Homo Oeconomicus, 34,* 259–273. https://doi.org/10.1007/s41412-017-0056-1

Black, F., & Scholes, M. (1973, May–June). The pricing of options and corporate liabilities. *The Journal of Political Economy, 81*(3), 637–654. Published by: The University of Chicago Press.

Black, M. (1937). Vagueness. An exercise in logical analysis. *Philosophy of Science, 4*(4), 427–455.

Bonissone, P. (1980). A fuzzy sets based linguistic approach: Theory and applications. In T. I. Oren, C. M. Shub, & P. F. Roth (Eds.), *Proceedings of the 1980 Winter Simulation Conference* (pp. 99–111).

Borgonovo, E., Capelli, V., Maccheroni, F., & Marinacci, M. (2017). Risk analysis and decision theory: A bridge. *European Journal of Operational Research, 264,* 280–293.

Borison, A. (2003, July). *Real options analysis: Where are the Emperor's clothes?* Presented at Real Options Conference, Washington, DC (pp. 1–30).

Box, G., Hunter, J., & Hunter, W. (2003). *Statistics for experimenters. Design, innovation, and discovery* (2nd ed.). Wiley-Interscience, Wiley.

Bratvold, R., Laughton, D., Enloe, T., Borison, A., & Begg, S. (2005). A critical comparison of real option valuation methods: Assumptions, applicability, mechanics, and recommendations. Society of Petroleum Engineers. In *Proceedings of the 2005 SPE Annual Technical Conference and Exhibition,* Dallas, Texas, USA, 9–12 October. Paper SPE-97011.

Buckley, J. (2006). *Fuzzy probability and statistics.* Springer.

Carlton, M., & Devore, J. (2017). *Probability with applications in engineering, science, and technology* (2nd ed.). Springer International Publishing.

Carmona, R. (2014). *Statistical analysis of financial data in R* (2nd ed.). Springer Science +Business Media.

Chander, M. (2019). *An introduction to fuzzy set theory and fuzzy logic* (2nd ed.). MV Learning.

Chung, K. (2001). *A course in probability theory* (3rd ed.). Academic Press.

Costa, A., Schiozer, D., & Poletto, C. (2006). Use of uncertainty analysis to improve production history matching and the decision-making process. *Society of Petroleum Engineers.* In *Proceeding of the SPE EUROPEC/EAGE Annual Conference and Exhibition,* Vienna, Austria, 12–15 June. Paper SPE-99324.

Cox, D. (2000). *The theory of the design of experiments.* Chapman & Hall/CRC.

Cox, L. (2015). *Breakthroughs in decision science and risk analysis.* Wiley.

Darwich, A., Hebert, P., Bigand, A., & Mohanna, Y. (2019). Background subtraction based on a new fuzzy mixture of Gaussians for moving object detection. *Journal of Imaging, 4,* 92. https://doi.org/10.3390/jimaging4070092

DasGupta, A. (2011). *Probability for statistics and machine learning. Fundamentals and advanced topics.* Springer Science +Business Media.

Davis, J. (2002). *Statistics and data analysis in geology* (3rd ed.). Wiley.

De Ville, D., Nachtegael, M., Van der Weken, D., Kerre, E., Philips, W., & Lemahieu, I. (2003, August). Noise reduction by fuzzy image filtering. *IEEE Transactions on Fuzzy Systems, 11*(4), 429–436.

Durrett, R. (2010). *Probability* (4th ed.). Cambridge University Press.

Edwards, W. (1954). The theory of decision making. *Psychological Bulletin, 51*(4), 380–417.

Etner, J., & Jeleva, M. (2012). Decision theory under ambiguity. *Journal of Economic Surveys, 26*(2), 234–270.

Fernandes, B., Cunha, J., & Ferreira, P. (2011). The use of real options approach in energy sector investments. *Renewable and Sustainable Energy Reviews, 15*, 4491–4497.

Fishburn, P. (1968, January). Utility theory. *Management Science, 14*(5), 335–378. Theory Series.

Fisher, R. (1922). On the mathematical foundations of theoretical statistics. *Philosophical Transactions of the Royal Society of London, Series A, 222*, 309–368.

Gaines, B. (1978). Fuzzy and probability uncertainty logics. *Information and Control, 38*, 154–169.

Guo, H., & Mettas, A. (2012, January). Design of experiments and data analysis. In *Proceedings of the 20212 Reliability and Maintainability Symposium* (pp. 1–11).

Hammond, J. (1967, November). Better decisions with preference theory. *Harvard Business Review, 45*(6), 123–141.

Howard, R. (1970, May). Decision analysis. Perspectives on inference, decision, and experimentation. *Proceedings of the IEEE, 58*(5), 632–642.

Howard, R. (1988, June). Decision analysis: Practice and promise. *Management Science, 34*(6), 679–695.

Hurwicz, L. (1951, February 8). The generalised Bayes Minimax principle: A criterion for decision making under uncertainty. *Cowles Commission Discussion Paper 355*, 1–7.

Jablonowski, C., Ramachandran, H., & Lasdon, L. (2011). Modelling facility-expansion options under uncertainty. *Society of Petroleum Engineers*, SPE Projects, Facilities & Construction, December. Paper SPE-134678-PA.

Jamshidi, A., Yazdani-Chamzini, A., Yakhchali, S., & Khaleghi, S. (2013). Developing a new fuzzy inference system for pipeline risk assessment. *Journal of Loss Prevention in the Process Industries, 26*, 197–208.

Kahneman, D., & Tversky, D. (1984). Choices, values, and frames. *American Psychologist, 39*(4), 341–350.

Kaltenbach, H. (2012). *A concise guide to statistics*. Springer Science +Business Media, Springer.

Karni, E. (2014). Axiomatic foundations of expected utility and subjective probability. In J. Mark, W. Machina, & K. Viscusi (Eds.), *Handbook of the economics of risk and uncertainty* (Vol. 1, pp. 1–39).

Keeney, R. (1982, Sep–Oct). Decision analysis: An overview. *Operations Research, 30*(5), 803–838.

Klir, G., & Yuan, B. (1995). *Fuzzy sets and fuzzy logic. Theory and applications*. Prentice Hall PTR.

Knight, F. (1921). *Risk, uncertainty and profit*. Houghton Mifflin Company, The Riverside Press.

Kobayashi, H., Mark, B., & Turin, W. (2012). *Probability, random processes, and statistical analysis*. Cambridge University Press.

Koch, K. (2007). *Introduction to Bayesian statistics* (2nd ed.). Springer.

Kosko, B. (1990). Fuzziness vs. probability. *International Journal of General Systems, 17*, 211–240.

Law, A., & Kelton, W. (1991). *Simulation modeling and analysis* (2nd ed.). McGraw-Hill, Inc.

Lawson, J. (2015). *Design and analysis of experiments with R*. CRC Press, Taylor & Francis Group.

Leslie, K., & Michaels, M. (1997). The real power of real options. *The McKinsey Quarterly, 3*, 4–23.

Li, B., & Babu, G. (2019). *A graduate course on statistical inference*. Springer Texts in Statistics, Springer Science +Business Media, LLC, Springer Nature.

Lima, G., & Suslick, S. (2005). An integration of real options and utility theory for evaluation and strategy decision-making in oil development and production projects. *Society of Petroleum Engineers*. In *Proceeding of the 2005 SPE Hydrocarbon Economics and Evaluation Symposium*, 3–5 April. Paper SPE-94665.

Malhotra, V., & Lee, M. (2004). Decisions and uncertainty management: Expertise matters. *Society of Petroleum Engineers*. In *Proceedings of the SPE Asia Pacific Oil and Gas Conference and Exhibition*, Perth, Australia, 18–20 October. Paper SPE-88511.

McNamee, P., & Celona, J. (2001). *Decision analysis for the professional* (4th ed.). SmartOrg Inc.

Mishra, S. (2002). *Assigning probability distributions to input parameters of performance assessment models*. INTERA Inc.

Mizumoto, M., & Tanaka, K. (1981). Fuzzy sets and their operations. *Information and Control, 48*, 30–48.

Moczydlower, B., Salamao, M., Branco, C., Romeu, R., Homen, T., Freitas, L., & Lima, H. (2012). Development of the Brazilian pre-salt fields-when to pay for information and when to pay for flexibility. *Society of Petroleum Engineers*. In *Proceeding of the SPE Latin American and Caribbean Petroleum Engineering Conference*, Mexico City, Mexico, 16–18 April. Paper SPE-152860.

Montgomery, D., & Runger, G. (2003). *Applied statistical and probability for engineers*. Wiley.

Murtha, J. (1997, April). Monte Carlo Simulation: Its status and future. *Society of Petroleum Engineers. Journal of Petroleum Technology*, 361–373.

Newendorp, P., & Campbell, J. (1971). *Expected value—A logic for decision making*. American Institute of Mining, Metallurgical and Petroleum Engineers. Paper SPE 3327.

North, D. (1968, September). A tutorial introduction to decision theory. *IEEE Transaction on Systems Science and Cybernetics*, SSC-4(3), 200–210.

Okuda, T., Tanaka, H., & Asai, K. (1978). A formulation of fuzzy decision problems with fuzzy information using probability measures of fuzzy events. *Information and Control, 38*, 135–147.

Olive, D. (2014). *Statistical theory and inference*. Springer International Publishing Switzerland.

Oliveira, J., & Baltazar, D. Evaluation of real options in an oil field. *Advances in Mathematical and Computational Methods*, 23–28.

Olofsson, P., & Andersson, M. (2012). *Statistics and stochastic processes* (2nd ed.). Wiley.

Patumona, Y. (2015, July). Real options method vs. discounted cash flow method to analyze upstream oil & gas projects. *PM World Journal, IV*(VII), 1–26.

Pratt, J. (1964, Jan–Apr). Risk aversion in the small and in the large. *Econometrica, 32*(1/2), 122–136.

Rabin, M. (2000, September). Risk aversion and expected utility theory. *Econometrica, 68*(5), 1281–1292.

Rasch, D., & Schott, D. (2018). *Mathematical statistics*. Wiley.

Raychaudhuri, S. (2008). Introduction to Monte Carlo simulation. In S. J. Mason, R. R. Hill, L. Monch, O. Rose, T, Jefferson, & J. W. Fowler (Eds.), *Proceedings of the 2008 Winter Simulation Conference*.

Ridza, P., Ya'acob, A., Zainol, N., & Mortan, S. (2020). Application of two level factorial design to study the microbe growth inhibition by pineapple leaves juice. *IOP Conference Series: Materials Science and Engineering, 736*, 022011. https://doi.org/10.1088/1757-899X/736/2/022011

Ross, Sh. (2009). *Probability and statistics for engineers and scientists* (4th ed.). Elsevier.

Ross, T. (2010). *Fuzzy logic with engineering applications* (3rd ed.). Wiley.

Sabri, N., Aljunid, S., Salim, M., Badlishah, R., Kamaruddin, R., & Malek, M. (2013, July). Fuzzy inference system: Short review and design. *International Review of Automatic Control, 6*(4), 441–449.

Sampath, S. (2001). *Sampling theory and methods*. Narosa Publishing House.

Santos, S., & Schiozer, D. (2017). Assessing the value of information according to attitudes towards downside risk and upside potential. *Society of Petroleum Engineers*. In *Proceeding of the SPE EUROPEC at 79th EAGE Conference and Exhibition*, Paris, France, 12–15 June. Paper SPE-185841-MS.

Sivanandam, S., Sumathi, S., & Deepa, S. (2007). *Introduction to fuzzy logic using MATLAB*. Springer, Springer Science + Media.

Smith, J., & McCardle, K. (1999). Options in the real world: Lessons learned in evaluating oil and gas investment. *Operations Research, 47*(1), 1–15.

Smith, J., & von Winterfeldt, D. (2004, May). Decision analysis in management science. *Management Science, 50*(5), 561–574.

Steagall, D., & Schiozer, D. (2001). Uncertainty analysis in reservoir production forecast during appraisal and pilot production phases. *Society of Petroleum Engineers.* In *Proceedings of the SPE Reservoir Simulation Symposium*, Houston, Texas, 11–14 February. Paper SPE-66399.

Soong, T. (2004). *Fundamentals of probability and statistics for engineers.* Wiley.

Strat, T. (1990). Decision analysis using belief functions. *International Journal of Approximate Reasoning, 4*, 391–417.

Suslick, S., Schiozer, D., & Rodriguez, M. (2009). Uncertainty and risk analysis in petroleum exploration and production. *TERRAE, 6*(1), 30–41.

Thomas, P. (2015). Measuring risk-aversion: The challenge. *Measurement, Elsevier, 79*, 285–301.

Thomopoulos, N. (2018). *Probability distributions with truncated.* Springer International Publishing, AG, part of Springer Nature.

Tversky, A., & Kahneman, D. (1981, January 30). The framing of decisions and the psychology of choice. *Science, 211*, 453–458.

Vahedi, A., Gorjy, F., Scarr, K., Sawiris, R., Singh, U., Montgomery, P., Clinch, S., & Sawiak, (2005). A. Generation of probabilistic reserves distributions from material balance models using an experimental design methodology. *Society of Petroleum Engineers.* In *International Petroleum Technology Conference*, Doha, Qatar, 21–23 November. Paper IPTC-11009.

Vilela, M., Oluyemi, G., & Petrovski, A. (2018). fuzzy data analysis methodology for the assessment of value of information in the oil and gas industry. In *Proceeding of the 2018 IEEE International Conference on Fuzzy Systems (FUZZ-IEEE)* (pp. 1540–1546).

Vinayagam, V. (2015). Integrating flexibility in new development to formulate optimal depletion strategy with reduced risk in a complex carbonate reservoir in Middle East. *Society of Petroleum Engineers.* In *Proceeding of the SPE Middle East Oil and Gas Show and Conference,* Manama, Bahrain, 8–11 March. Paper SPE-172606-MS.

Wackerly, D., Mendenhall, W., & Scheaffer, R. (2008). *Mathematical statistics with applications* (7th ed.). Thomson Learning Inc.

Whalen, T., & Bronn, C. (1988). Essentials of decision making under generalized uncertainty. In J. Kacprzyk & M. Fedrizzi (Eds.), *Combining fuzzy imprecision with probabilistic uncertainty in decision making* (pp. 1–11). Springer.

Wilson, E. (2015). A practical guide to value of information analysis. *Pharmaco Economics, Springer International Publishing, 33*, 105–121.

Yoe, Ch. (2012). *Principles of risk analysis.* CRC Press, Taylor & Francis Group.

Zhao, T., & Tseng, Ch. (2003, September). Valuing flexibility in infrastructure expansion. *Journal of Infrastructure Systems*, 89–97.

Zimmermann, H. (1996). *Fuzzy set theory and its applications.* Kluwer Academics Publishers.

Printed in the United States
by Baker & Taylor Publisher Services